An Open Pit Visible from the Moon

The Environment in Modern North America

Leisl Carr Childers and Michael Childers, Series Editors

An Open Pit Visible from the Moon

The Wilderness Act and the Fight to Protect Miners Ridge and the Public Interest

ADAM M. SOWARDS

UNIVERSITY OF OKLAHOMA PRESS : NORMAN

Library of Congress Cataloging-in-Publication Data

Names: Sowards, Adam M., author.
Title: An open pit visible from the moon : the Wilderness Act and the fight to protect Miners Ridge and the public interest / Adam M. Sowards.
Description: Norman : University of Oklahoma Press, 2020. | Series: Environment in modern North America | Includes bibliographical references and index. | Summary: "Combines rigorous analysis and deft storytelling of the struggle to define the contours of the Wilderness Act when the Kennecott Copper Corporation announced its plan to develop an open-pit mine in the Glacier Peak Wilderness Area"—Provided by publisher.
Identifiers: LCCN 2019049766 | ISBN 978-0-8061-6501-1 (hardcover)
Subjects: LCSH: Environmentalism—Political aspects—United States. | Copper mines and mining—Environmental aspects—Washington (State)—Glacier Peak Wilderness | Strip-mining—Environmental aspects—Washington (State)—Glacier Peak Wilderness | United States. Wilderness Act. | Kennecott Copper Corporation. | Public interest—United States. | Glacier Peak Wilderness (Wash.)—Environmental conditions.
Classification: LCC GE197 .S68 2020 | DDC 333.78/216097977—dc23
LC record available at https://lccn.loc.gov/2019049766

An Open Pit Visible from the Moon: The Wilderness Act and the Fight to Protect Miners Ridge and the Public Interest is Volume 2 in the Environment in Modern North America series.

The paper in this book meets the guidelines for permanence and durability of the Committee on Production Guidelines for Book Longevity of the Council on Library Resources, Inc. ∞

For Mom and Dad,
who created a home, in the shadow of the Cascades,
where I could flourish

CONTENTS

MAPS

ACKNOWLEDGMENTS

SINCE WRITING THIS BOOK, I have often joked that I wrote it by accident, but I never joked that I did it alone. I am grateful to be able to publicly acknowledge the support I received along the way.

Jeff Crane invited me to contribute an essay to a book he and Char Miller were editing on grassroots environmentalism. I resisted, insisting that I had nothing to offer. Jeff was relentless, so I suggested that I write about this wilderness campaign, thinking it would be a short, quick, one-off essay before I moved on to another book project. But when I dove into the research, I realized this story deserved its own book; four years later, here it is. My initial thanks, then, go to Jeff and Char, who prompted this project and whose enthusiasm sustained it at the outset. That first essay appeared in *The Nature of Hope: Grassroots Organizing, Environmental Justice, and Political Change,* published by the University Press of Colorado, which graciously provided permission for me to use material first appearing there.

Writing a book depends on many practical matters. Archivists do the work that allows historians to research and write. Staff at the Bancroft Library at the University of California, Berkeley; Special Collections at the University of Washington Libraries; and the National Archives at Seattle were essential to my research. Samantha Richert of the National Park Service unearthed some important photographs for me. My home institution, the University of Idaho, also facilitated this project. The College of Letters, Arts, and Social Sciences provided research funding that allowed me to travel and focus. Former college dean Andy Kersten created these grant and fellowship opportunities, and my former department chair Sean Quinlan supported my applications for them. The department also helped defray the costs of the maps and photos. My sincere thanks to all who helped establish this foundation.

Although my methodology focused on archival and published materials, I connected with a handful of people who were involved in the North Cascades campaign, related organizations, or the wilderness movement. Those contacts improved my book and, even more importantly, inspired me by their commitment and generosity. Robert Michael Pyle, with his characteristic zestful spirit, shared his experience at the protest recounted in chapter 7. Philip Fenner, current president of the board of the North Cascades Conservation Council, not only granted me permission to use material from the organization's newsletters but also shared his own photographs of Miners Ridge and Glacier Peak. Longtime activist Doug Scott, who possesses a keen interest in wilderness history, answered my out-of-the-blue email and shared a vital perspective not available in the archives. At an unrelated meeting, I met Gary Paull, who worked on recreation for the U.S. Forest Service out of Darrington; he shared maps with me and saved me from making some mistakes regarding the land's legal status. If I have misconstrued anything from these helpful people, the errors are mine alone.

I shared my in-progress research on two memorable occasions. In 2017 Steve Schulte invited me to be the Wayne Aspinall Distinguished Chair in History and Political Science at Colorado Mesa University. I was grateful for that opportunity to present my initial conclusions and to ponder the good questions raised there that forced me to think harder about my arguments. Steve also helped me understand Representative Aspinall in crucial ways. In 2018 Crawford Gribben of Queen's University Belfast invited me to Northern Ireland to discuss my work with Northern Bridge Consortium postgraduate students, and presenting there was unforgettable experience that afforded me the distance I needed to see my work in a broader context.

I have the great fortune to be publishing this book with the University of Oklahoma Press. Now retired, Chuck Rankin acquired the book for the press. His initial interest and a critical reading of several chapters confirmed that I was on to something. Adam Kane took over as editor-in-chief and has been a model editor, providing the right amount of prodding and support. Steven Baker ushered in the final product with professionalism. All three always made me feel in exceptionally competent hands. The press connected me to Erin Greb as cartographer and Chris Dodge as copy editor, and both improved the book in incalculable ways. The press also ensured that the book proposal and the completed manuscript received excellent reviewers whose questions and insights improved my understanding, ideas, and expression. Initially this book was targeted for a series edited by Mark Fiege, and his interest in it helped propel me. However, it became clear that the manuscript fit better in the Environment in Modern North America series, edited by Leisl Carr Childers and Michael Childers. I am delighted to be part of this series and to have worked with Leisl and Mike, model scholars and editors I admire. They have championed this project more than a writer dares to expect. I hope this finished book rewards their faith in it.

In my experience, environmental historians have always been a welcoming audience and collegial group. I have benefited from innumerable conversations about my research at meetings of the American Society for Environmental History. I would not be able to list all those scholars whose work and whose conversations have shaped my work, so I have to be content to highlight a few and apologize to those omitted. Mark Harvey and Lauren Danner both shared archival material with me from their own research. Steve Pyne read an early version of the essay that appeared in *The Nature of Hope* and provided trenchant literary advice (that I should probably have taken more to heart) and continued to be the sort of mentor I have appreciated since graduate school. Kevin Marsh and Jay Turner remain my wilderness history teachers, and I relied on their scholarship and friendly answers whenever I queried them. Similarly, I drew on Jamie Skillen's knowledge of public lands history to strengthen the manuscript. And everyone should have a friend and colleague like Lisa Brady, who has been a steadfast supporter and confidante during this project (and long before); I hope she realizes how much I admire and rely on her good judgment and friendship.

Living in the same town as another Northwest environmental historian has distinct advantages. Jeff Sanders deserves special thanks. He read the manuscript in full and in pieces, over and over. His critical eye saved me from mistakes and embarrassment countless times. His incisive questions pushed me further in my thinking and pointed out oversights. Our regular meetings over coffee and walks around town have become a type of intellectual and personal sustenance for me that I never expected. For more than a decade now, we have grown as scholars, writers, and people together, and Jeff has become a most important influence. I hope I can repay his assistance and friendship in kind.

No one matters to this process more than my spouse, Kelley. Her faith and support in what I do is boundless and sometimes a substitute for my own. Her patience with the inevitable ups and downs writing a book entails means that she deserves rewards and thanks more than this acknowledgement can provide. With endless openness she listened to my discoveries, shared in my ideas, and read my words—always with enthusiasm and charity. For the last decade, we have shared a full, loving life together that reminds me every day what matters most.

I dedicate this book to my parents, Ruth and Howard Sowards. I grew up in a house whose living room looked out on the Cascade Mountains, a constant presence then and now. I grew up in a home where my parents' love and support were also a constant presence, without which I would have been unable to accomplish what I have or be who I am. A book dedication is not enough, but I hope they know it represents my appreciation for what they have given me and all my love for who they are.

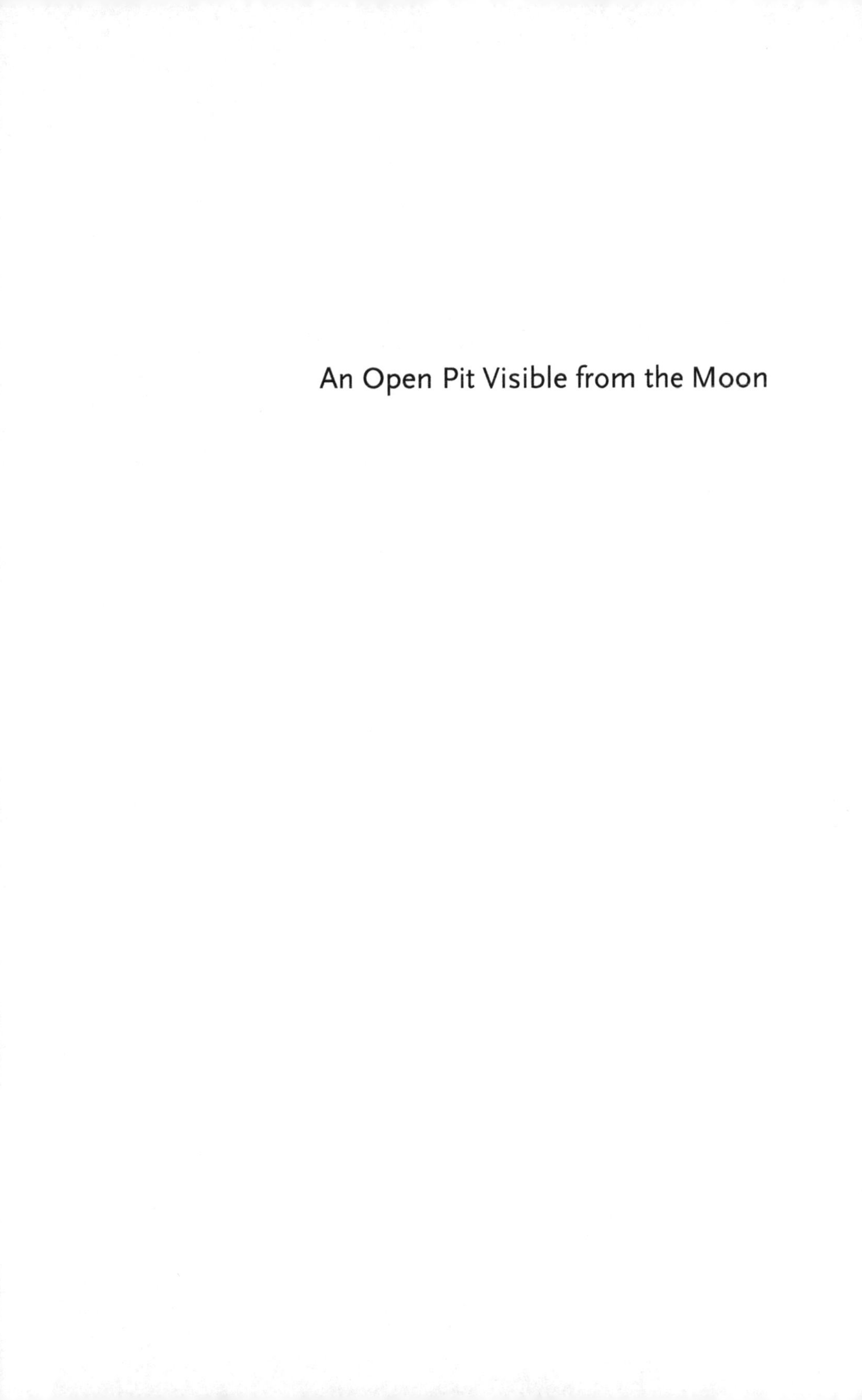

An Open Pit Visible from the Moon

Introduction: The Telescope

BIG, BOLD LETTERS FILLED nearly a third of a page:

"An Open Pit,
Big Enough
To Be Seen
From The Moon"

Beneath them, centered on the page, a photograph of an open-pit mine drew readers' attention to a desolate landscape, where distant mountains gave way to a hole descending deep into the earth, a stone whirlpool lacking any sign of life. An engineering marvel, perhaps, but clearly not wild like the bumpy ridges and plunging valleys rolling toward the horizon behind the pit. Its unnaturalness suggested something sinister, even diabolical. Given that this was the late 1960s, viewers might have taken it for a bomb crater, although it was larger and more neatly terraced than ones televised from Vietnam. The accompanying article offered no reassurance, telling of a profiteering company that planned to excavate a similar hole—nearly seven football fields wide and deep enough to contain Seattle's Space Needle. During excavation, Kennecott Copper Corporation would dump more than twenty-six million pounds of waste rock in the middle of Washington State's Glacier Peak Wilderness Area.[1]

In this wilderness, the eponymous peak dominated, but Plummer Mountain, which lay just to the northeast of the snow-topped volcano, contained the copper. It also drew visitors. At the end of the 1960s, several men hiked up the mountain, and one later described the view of "broad alpine meadows interspersed with strands of spruce." Of a spot known as the Golf Course to those who explored nearby for minerals, the writer said, with sly hyperbole, "With very little bulldozing, [it] could in fact become one of the seven wonders of sport, with the red wall of Plummer Mountain above it, the deep valley of the Suiattle [River] falling away beside it, and the sparkling, spectacular imminence of Glacier Peak in full view from every tee, fairway, and green."[2] Here, amid spectacular splendor, Kennecott Copper wanted to dig its gigantic pit, a sacrilege to those who revered wild places.

One of the reverent hikers was David Brower, the executive director of the Sierra Club, arguably the nation's most important conservation organization. Brower was known to place rousing political ads in newspapers, like the one described above. In fact, his ads attracted the attention of the Internal Revenue Service, and subsequent investigations ultimately led the IRS to revoke the club's tax-exempt status.[3] In the alpine meadows around Plummer Mountain, as the hikers investigated rocks for copper, Brower mused about the "man-made crater" that would be "so large it would be visible from the moon." The writer on this hike, the *New Yorker*'s John McPhee, pointed out that the claim was exaggerated, and Brower admitted that seeing the mine from the moon would require "a small telescope."[4]

The claim about the mine's magnitude and the ad that spread it became folklore, part of the story widely told of conservationists confronting a copper company in the Cascade Mountains to keep wilderness wild.[5] Those who remember Kennecott Copper's claim in the Glacier Peak wilderness share a common narrative: Kennecott had a legitimate stake and planned to construct a significant mining operation, but conservationists fought back and stopped the open-pit mine with scrappy activism that produced sufficient political pressure to halt the mighty company.[6] As with the tantalizing ad and Brower's exaggeration, that outline is true enough in broad form. This book, like Brower's small telescope, provides the clarifying vision necessary to see the details. When magnified and investigated, the history in focus is still inspiring but also complicated, contingent, and unexpected. It is no simple morality play.

Histories of wilderness threatened and preserved often take on a Manichean cast, with heroes and villains filling roles preordained. In this history, nothing and no one was foreordained; why it ended as it did was not predictable. In 1964 the company possessed every legal right to mine in wilderness, and there was no reason to think that by 1970 trucks would not be hauling ore down the mountain to a smelter in Tacoma. Recapturing not only this sense of urgency but also the uncertainty of the outcome reminds us that campaigns for wilderness—or endangered species or decarbonization or civil rights—do not bend toward an endpoint everyone can see from the outset. Resolutions are reached by people deciding to act, by people making history, against a backdrop of forces well beyond individuals' ability to fully imagine, see, or control. This enterprise of mining mountains included a complicated amalgam of geography, law, politics, economics, and activism. People traveled to this fight along many paths, like tributaries joining a river. As with any political campaign, the diversity of people and factors involved multiplied the possible outcomes. Stories of simple good and evil, hero and villain, justice prevailing or thwarted, entail unrealistic caricatures. And to know the on-the-ground struggle for wilderness is to reckon with the very ground itself.

Stories have settings, the places where the drama unfolds. Environmental history as a field of study has reminded historians that history, too, develops in place.[7] The North Cascades provide the ground where this story trends from tragedy to comedy. The Cascade Mountains at their southern and northern extremes lean over the California-Oregon and the British Columbia–Washington borders, but they primarily run the length of the two Pacific Northwest states. The North Cascades proper, from Stevens Pass to the Canadian border, occupy roughly one hundred square miles of wild, tangled mountains and river valleys.[8] Two snowy volcanoes, Mount Baker and Glacier Peak, push up more than ten thousand feet in elevation, standing as sentinels amid countless sharply rising and falling mountains and drainages, the evidence of tectonic plates crashing into each other, volcanism forcing rocks upward, and glaciers carving their way through millennia. The mountains block the moisture-filled clouds that move from the Pacific Ocean. To relieve their load on the prevailing eastward journey, clouds drop snow and rain heavily along the western slopes, feeding rushing rivers and helping to produce the thick Douglas fir forests characteristic of the coastal Northwest. The Cascades' eastern slopes are consequently much drier, but river valleys still fill with pine forests. Visitors recognize the North Cascades by their familiar high ridgelines and deep valleys, sharper and wilder than most western mountains. As a setting, the North Cascades provide rugged, jaw-dropping beauty suitable for an inspiring story.[9]

Within the North Cascades, this story focuses on a specific place, Miners Ridge.[10] Centered at what many have called the "scenic climax" of the region, Miners Ridge runs east–west in the Suiattle River drainage, just to the west of the Cascade crest at Suiattle Pass and seven or eight miles northeast of the Glacier Peak summit.[11] Plummer Mountain marks the ridge's eastern edge, while its western terminus is squeezed where Canyon Creek swings from the north and pinches the ridge into the Suiattle River. Backpackers from the greater Seattle area, seventy-five air miles to the southwest, typically start their trip from the end of the Suiattle River Road, hiking the low and often damp ground through towering and moss-covered forests near the river for close to ten miles while barely rising in elevation. Then trekkers must ascend, switch-backing up through the trees before emerging above timberline, ultimately forty-five hundred feet higher than the trailhead and fifteen or so miles beyond it. Backpackers can connect here with the Pacific Crest Trail and head north to Canada or south to Mexico, or they can continue east through Suiattle Pass and on down to the majestic Lake Chelan, a glacier-carved lake with depths of nearly fifteen hundred feet that has long been a vacation destination for its quiet beauty.

Back on the ridge, today's hikers, just like Brower and McPhee, lose their breath not only from the climb but also from the view. A look back over the Suiattle River Valley's dense green forests offers stunning vistas, but that pales compared to Glacier Peak that dominates the southern views with its hulking white conical mass. Just below the ridge, Image Lake attracts many who have sought mountain transcendence. This mountain tarn with glacial origins is a mere three acres but is situated to capture Glacier Peak's glory and reflect it almost perfectly. Clear days have provided countless trekkers with phenomenal views and photographs. On Miners Ridge, after the timber thins out, wildflower meadows add a soft touch to the slopes. And here above timberline, just beyond the lake that provides iconic Northwest images, lies rock impregnated with copper.[12] In this place, a few hundred acres on a ridge among countless ridges, the story unfolds.

As the history presented in this book suggests, however, place is more than just a physical setting. Place is complicated and complex. The specificity of Miners Ridge, with its unique material properties, is central to this environmental history.[13] Without the geological forces that created the copper, there would have been no temptation to mine and no controversy to investigate. And without the particularly beautiful scene that spread out from Miners Ridge, fewer conservationists would have fought as hard as they did to protect it. This discrete place, like all places, is connected to many others beyond it. In Washington, D.C., Congress and the U.S. Forest Service made defining policy decisions. In New York City, Kennecott Copper Corporation weighed profits and losses. In Seattle, conservationists sent champions to the mountains and roused the public with their tales. Policy, profit, and public interest entangled these distant places, binding them together. While a chief executive or chief forester might be expected in a story about a mine in a national forest, other factors were also important in this narrative, such as labor strikes in Chile, the war in Vietnam, and prices at the London Metal Exchange. This ridgeline in eastern Snohomish County of Washington State, then, interacted with a wide world. Many visited Glacier Peak wilderness to escape the world, to be connected to the rejuvenating scent of wildflowers, the annoying buzz of biting flies, and the bracing chill of the west wind. But this wilderness was no escape, not outside history or time; it was connected to diplomatic and economic decisions from half a world away. The Cold War, for example, generated the very prosperity that allowed citizens to be secure enough financially to debate and define the public good through wilderness.

Miners Ridge was wilderness in both abstract and concrete ways. The ridgeline jutted up, standing twenty-five miles from any permanent human population and more than seventy-five air miles from any city. No dam plugged up Miners Creek. No road carried trucks to Image Lake. No fences penned in cattle or sheep. Its distant ruggedness kept it inaccessible, and that inaccessibility kept it wild, something of increasing value after industrialization and urbanization ran riot across the wild and rural countryside. Nineteenth-century Romantics began to see spiritual value and aesthetic beauty in wild mountains, where they once had felt only fear.[14] By the twentieth century, urban Americans, especially elite professionals—writers, doctors, attorneys, planners—had developed a strong culture around wilderness and urged protecting wild places from ecologically damaging activities and excessive commercialism. They asserted that wilderness served several distinct but connected roles, including as a place to find traces of the divine, test one's physical limits, escape unhealthy cities, study wildlife, and preserve frontier conditions as historical and cultural reference points. Altogether these thinkers saw the wild as an antidote to the social and psychological problems they witnessed rising in American cities. Meanwhile, suburbs expanded and residents headed to mountains with Boy Scout troops and on family camping trips, generating memorable experiences and raising a generation of children who took to the streets to protect the trails.[15]

Most of those roles for wilderness were also broadly intellectual goals and thus abstract, philosophical, or otherwise ethereal. But to ensure that wilderness retained its essential qualities and remained protected required government action, because most places that seemed wild were part of the public domain, such as western mountains like the Cascades. That made wilderness a political question. And politics demanded concrete actions more than pondering philosophies. When the U.S. Forest Service started managing lands for wilderness values in the 1920s, it did so while paying close attention to specific places; when conservationists started challenging the Forest Service's management of wilderness, they did so while paying close attention to individual locales.[16] Although intellectuals may have developed general principles, battles about wilderness occurred over precise places on the map and in the woods, mountains, and deserts. To adapt a phrase, democracy sits in places—places like Miners Ridge.[17]

The story of Miners Ridge reveals a critical episode in wilderness history that has until now been treated mostly as a footnote.[18] It shows how wilderness was made, challenged, and championed in a crucible when the law and concept remained malleable, with Americans still struggling to define its limits and protect its integrity. Passing the Wilderness Act in 1964 was an impressive achievement for conservationists. The law provided for the strictest land-use regulations in the nation. It strengthened wilderness protection by vesting it in Congress rather than in the Forest Service where changes were easier to enact and prevented commercial activity across millions of acres. The Wilderness Act traveled winding legislative routes, becoming law only after a series of compromises. Howard Zahniser, who initially drew up the law, was the architect, but Wayne Aspinall was the engineer who made it happen. Zahniser, a modest and hardworking man, served as the executive secretary of the Wilderness Society and drafted the first bill in 1956. Until his death in 1964, Zahniser met with countless politicians, administrators, and industry representatives—"the constant advocate," his biographer called him—trying to convince them to support the wilderness bill that he revised dozens of times to satisfy their concerns.[19] Aspinall, a dedicated and independent-minded legislator from western Colorado, chaired the House of Representatives Interior and Insular Affairs Committee, through which all wilderness bills had to pass. He was committed to multiple-use development on public lands and practiced legislative delay to ensure western commodity powers, mainly ranchers and mining companies, an opportunity to strategize and protect their interests. Zahniser and conservationists had no choice but to compromise with Aspinall if they wanted their bill out of his committee, a prerequisite for it to become a law. So they gritted their teeth and agreed to a provision that allowed prospecting in designated wilderness areas to continue for twenty years, and they could not prevent developing mines and processing plants on bona fide claims.[20] This mining exception made what happened at Miners Ridge important, controversial, and, above all, historic.

The copper presented everyone with a test. Kennecott Copper Corporation needed to establish its right to use the Wilderness Act's mining exception, for its own interests at Miners Ridge and elsewhere, and the mining industry at large followed this episode with the keenest interest. For their part, conservationists hoped to beat back the industry, show that wilderness and mining were incompatible inherently, and effectively void the Wilderness Act's mining provision. To be sure, the Glacier Peak Wilderness Area mattered greatly to conservationists, especially those based in the Pacific Northwest, but conservationists generally wanted to establish a precedent that might prevent mining in *any* wilderness. Meanwhile, the Forest Service used this campaign to assert both its bureaucratic power to regulate mining and its longstanding history of protecting wilderness values, something the agency insisted repeatedly to conservationists' disbelief. Through it all, politicians felt frustrated if not confused. They passed the Wilderness Act, thinking they had settled the question, yet the public kept arguing. Legislators had written the test and created the answer key, but the public now disputed the score. Policy scholars sometimes refer to path dependency to explain how it becomes difficult to change course once a path is set.[21] All parties sought to blaze the way, to shape how the Wilderness Act would be implemented, because they understood that these early years were the critical time to establish lasting direction.

Miners Ridge tested the loophole's durability, the most prominent example where a company invoked the mining exception. How corporations, conservationists, agencies, politicians, and citizens responded shows us the nature of political power as it was being defined, thrusted, and parried in and for public space. The visions clashing at Miners Ridge informed what would happen in the Sierra Nevada or the Rocky Mountains or anywhere else wilderness hid valuable minerals. This book explains the test results.[22]

Today no mine mars Miners Ridge, a fact that suggests conservationists won. And they did—at least in the specific sense that Kennecott did not build the mine and no longer holds the title to the land. For activists or backpackers who aimed to prevent open-pit desecration on Miners Ridge, that test result may suffice. Knowing the result, however, may not be as revealing or as important as understanding the process by which it occurred, because the ways that activists pushed the corporation and politicians, the way writers instructed their readers, and the way the public rallied to call for protection all showed the power of citizenship. And the meaning that can be gleaned from the stories made in the Cascades connects this place, conflict, and people to broader forces and concerns. For historians who want to know the fuller story—and deeper implications—more is needed. The simplest history examines Miners Ridge in the North Cascades and whether it would be mined. But the account I offer goes beyond that in investigating the implementation of the Wilderness Act—its limitations and possibilities—across millions of acres of the United States nowhere near Glacier Peak Wilderness Area. This was a critical topic after 1964. It also is concerned with the public responsibilities of private companies, a timeless topic for any society and government to consider. At its core, this book helps us consider democratic power and the public interest: the power of politics and law, the power of corporate influences and priorities, the power of the public to question and challenge those powers.

It also is, I hope, a book of stories, good stories about interesting people and beautiful places that matter. The book comprises three parts. In the first, "Bedrock," I explain the powerful institutions that occupied the North Cascades and set the trajectory that resulted in the standoff of the mid to late 1960s. The U.S. Forest Service played the largest role, as it sought to manage forests, promote multiple uses, and protect wilderness. Like many a government agency, the Forest Service also worked as a broker between competing interests, including Kennecott Copper and other mining concerns, local conservationists that gathered strength over the first two-thirds of the century, and the National Park Service, which coveted the North Cascades and had its own designs for the region. Policies and proximity put all these forces and factors together in a context where laws and traditions, forged in public and through time, tested their ability to achieve their incompatible goals for Miners Ridge and the larger North Cascades landscape.

The second part, "Challenges," the heart of the book, recounts a series of events and people who challenged Kennecott's plans to develop its property, a long-rumored possibility but finally presented in late 1966. The stories in this section show the ways, often clever and unique, that conservationists confronted the company. There are meetings in corporate boardrooms, to be sure, but there are also tales of a Supreme Court justice inspiring college students along a trail, a local doctor disrupting a shareholder meeting in a historic New York City hotel, and a college student who petitioned a corporate president (who then agreed to a meeting but became so nervous about it that he had to rehearse). All these efforts found people and groups asserting that the public possessed a legitimate interest in Miners Ridge, a public interest so powerful that it deserved to restrict corporate activities. This perspective extended democratic action in a challenge to private, corporate power and, at other times, government prerogatives. The campaigns detailed here all targeted Kennecott specifically, but they siphoned off from the broader historical currents of the 1960s that circumscribed power broadly.

The final part, "Resolution," wraps up the story, identifying the reasons the mine did not open and explaining how this campaign became part of the national environmental movement, a synecdoche of sorts for its significance. The narrative shifts here, since the story of Miners Ridge follows a slightly atypical pattern. From a distance, the story looks like dozens of other environmental histories: a beautiful place, insufficient protection, the threat of extraction, a rousing defense by the public, the threat fading, the place remaining beautiful. After my research, digging through countless archival collections, newspapers, and books, I have come to think that something both simpler and more complicated was at work. Economics explains the lack of the mine as well as any single factor—copper prices never got high enough for Kennecott. But price was always only one factor among many. You need to turn the pages to discover the combination of things that nudged history along the trail it took. Ultimately, unexpected turns ruined the narrative arc I anticipated constructing, but that makes this story more interesting and significant, not less. It exemplifies contingency, a concept that historians use often.[23] Historians emphasize it to demonstrate the various ways unpredictability shapes history's trajectory. Learning to recognize and appreciate contingency helps us understand historical agency better. In the battles over public lands and wilderness today, we would be wise to remember that contingency still reigns; the future is ours to *try* to construct.

I hope this history of the North Cascades, of a short-lived campaign with long-lasting significance, prompts us to think about actions withheld and choices made. In important ways, this history is a chronicle of things that did not happen. But because something did not happen does not make it unimportant. Just as a fire suppressed can be as ecologically significant as one that burns, plans that fail to materialize shape landscapes and history too.[24] Miners Ridge bears this out—and not just in the obvious way that a gigantic hole does not greet Pacific Crest Trail hikers as they move north from Glacier Peak. For example, part of this story is about North Cascades National Park, even though Miners Ridge is not part of the park. Working from today's park boundaries, a researcher might logically ignore Miners Ridge and miss its substantial influence. This, of course, is why historians do not work backward but seek to recount how history unfolded on its own terms, from the past toward the future. That philosophical and methodological perspective can surprise us, can move us to rethink our expectations, can push us to unsettle certainties—just like those conservationists half a century ago tried to unseat the inevitability of corporate power and legislative sanction.

Despite the love affair Americans have with private property and capitalism, grand and wild places remain protected from commercialism. That is the remarkable achievement and incongruity that wilderness, as well as anything, represents in U.S. history. In spots like Miners Ridge, places with views that pull you up short, where you gaze across panoramas and see across time, where your longing and imagination unwind without limits—in places like that, nature's mystery provokes. But what we see from atop Miners Ridge, when we awake to catch Glacier Peak mirrored in Image Lake, is not only nature untrammeled but also human choices across decades and centuries. Wilderness exists today because of choices people made and continue to make. The Miners Ridge controversy stirred something undeniable in the public. Tracing its history opens vistas for us consider and promote the public interest.

PART ONE
Bedrock

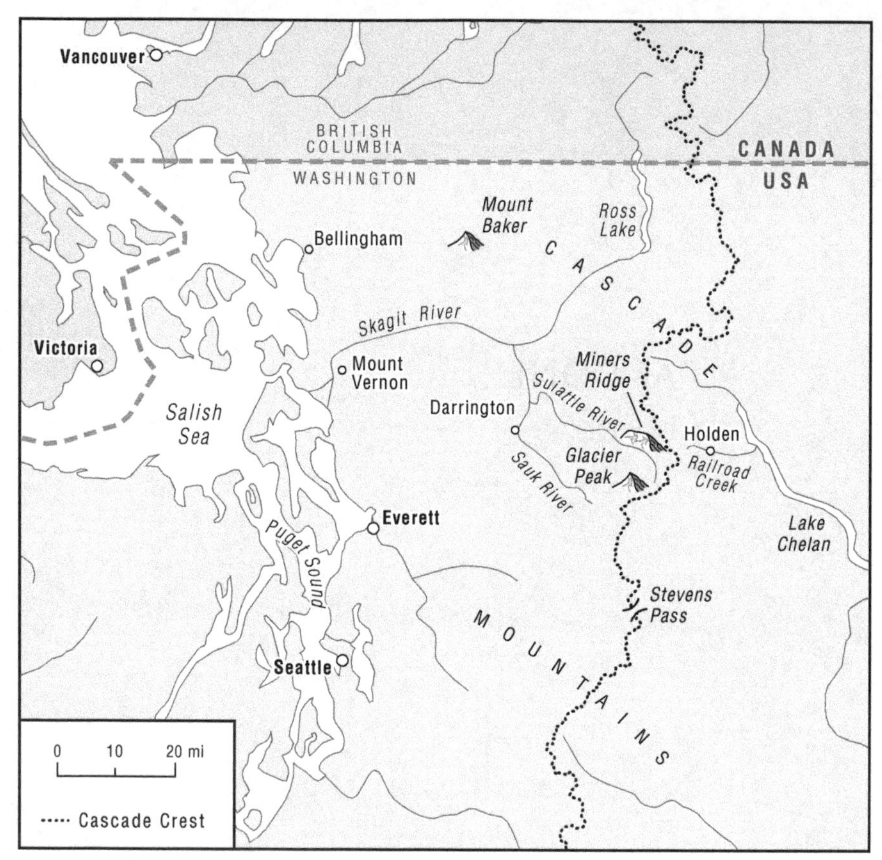

Major features of the North Cascade region

ONE | The Foundation

HISTORY BARELY REMEMBERS FRED CLEATOR, a recreation specialist and land planner for Region 6 of the U.S. Forest Service (USFS).[1] But in 1936 Cleator produced a noteworthy report that celebrated the agency's young history with wilderness preservation and expressed concern for the future of wilderness in the region. Although the Forest Service was formed in 1905 to secure the nation's future timber supply and protect watersheds, it had added other duties and enlarged its mission. By the 1920s the Forest Service recognized a distinct and leading role it might play in outdoor recreation, a stance it consciously promoted in part to counter the National Park Service's preeminence in that category. Cleator's report stands out for clearly articulating the agency's wilderness values and the way relatively unknown people shaped this history.

After traveling extensively in the North Cascades, Cleator recognized the area's wilderness appeal and the need for the agency to guard it. His "Report on Glacier Peak Wilderness Area" responded to the "strong demand" the Forest Service heard from people of the Northwest that the Glacier Peak region "should be left as nearly as possible in its natural state." Already the Forest Service had created a North Cascades Primitive Area snugged up to the Canadian border, and Cleator was charged with determining if the more southern Glacier Peak area deserved similar protection. In his estimation, it did. Cleator reported that the region was "naturally very inaccessible" and "exceedingly rugged." These qualities had kept the North Cascades relatively free from large timber, mining, and grazing operations that crept into many western mountain ranges and degraded wilderness qualities. "Since this country is inherently forbidding and most difficult of access," Cleator explained, "it is believed best suited for the purpose [of wilderness preservation], and least subject to possible future changes. It will tend to remove the temptation and feverishness of various future organizations and administrations for constant highway and roadway development of out of the way places which should be left alone."[2] With that, Cleator began his report and recommendations, aiming squarely for stability for Glacier Peak wilderness country.

His "Report on Glacier Peak Wilderness Area" made plain the Forest Service's perspective. As a federal bureaucrat, Cleator hoped to establish a policy for longevity, something often at odds with political or market cycles. He saw the Glacier Peak region threatened most not from traditional extractive industries but from roads, the central target of wilderness advocates organizing at this time.[3] If built, these ribbons of gravel and pavement, conveyors of cars and trucks, would tie this natural area to urban and industrial forces. Commercializing the backcountry was anathema here. Cleator observed the historical moment and witnessed frenzy. He scouted for a future that minimized uncertainty and maximized inaccessibility for wilderness. Cleator saw an opportunity for the Forest Service to outflank the National Park Service, then casting covetous eyes toward Cascade peaks and imagining a national park to capture the high country throughout Washington State. But in the 1930s, forceful figures in the Forest Service, led by Robert Marshall, with whom Cleator shared a memorable hike, saw wilderness recreation as a key part of the agency's future. Marshall and others argued that NPS road-building had ruined too many mountain retreats, and they were concerned about roads planned for the recently proposed Ice Peaks National Park along Washington's Cascade crest.[4] This moment saw a convergence of threats to wilderness: traditional extraction pressures, government conservation rivalries, and road-building momentum. Cleator, Marshall, and local conservationists looked to Glacier Peak country as a critical bulwark against machines invading the North Cascades and wondered what power they had to forestall them.

Although much of this book focuses on the looming threat of an industrial open-pit mine, Cleator's report alerts us to other contexts, long-standing in the region. Government agencies competed with one another. Factions *within* agencies clashed. Citizens pressed from outside government. Recreation preferences—motorized or not—rivaled each other in imagining how best to experience and enjoy the scenic treasures in the high Cascades. And roads, mines, and timber harvesting all attracted concern that bubbled out of local conservation groups, federal agencies, and national policies through the long crucible of this history. The "Report on Glacier Peak Wilderness Area" moved through bureaucratic channels midstream, as history flowed from the past toward the future. Cleator wrote it decades before Kennecott Copper Corporation acquired mining claims at Miners Ridge, but even then forces were in play. The first tests over who would shape the future of this landscape had arrived.

Today Miners Ridge is contained in the Glacier Peak Wilderness Area in the Mt. Baker–Snoqualmie National Forest as part of the Cascades Mountains of the

Pacific Northwest in U.S. Forest Service Region 6, just south of North Cascades National Park. It has not always been so. The ridge has not moved, but each of these places, every bureaucratic home, represents political and administrative choices wrought over time and typically contested. Policymakers and administrators changed designations at times to reflect evolving local and national priorities. Although they were sometimes ambiguous, these new designations symbolized different purposes, different levels of protection, and different categories of conservation. Excavating the history of Miners Ridge, just like digging up its ore, demands paying attention to intersecting layers, where the ecological and the political amalgamate.

Ecologically, Glacier Peak—which Sauk-Suiattle people know as Da Kobad, or Great White Mother Mountain—represents the North Cascades landscape as well as anything.[5] The hulking volcano towers over deep-cut valleys with roiling rivers. The White Chuck River gathers water off the peak's southwest side and grows before dumping into the Sauk River, which then moves west before turning sharply north, after which it swallows the Suiattle River. The Suiattle's own origins are on the southeast side of Glacier Peak, from which it wheels all the way around the mountain counterclockwise and drains a parallel valley north of the White Chuck and the south-facing slopes of Miners Ridge before entering the Sauk, which eventually joins the Skagit River on its way to emptying in the Salish Sea about fifty miles south of the U.S.-Canadian border. The mass of mountains from Stevens Pass north to the border contains countless streams, valleys, ranges, and peaks. West of the Cascade crest, jagged peaks like the Pickett Range rise high and have long enticed mountain climbers and awed backpackers. In the east slope's shadows, valleys containing rivers and the long, snaking Lake Chelan ultimately pour into and link the Northwest's two most significant natural features: the Cascade Mountains and the Columbia River.[6]

The natural setting established parameters for human history, encouraging mobility among distinct groups. Indigenous people who centered much of their time in this area, such as the Sauk-Suiattle people, moved freely through the broader landscape, from Puget Sound, up river valleys, and over the Cascade Mountains in a seasonal round that met their cultural and economic needs with salmon, berries, camas, and more. This mobility was not aimless wandering as old histories often portrayed but rather movement between specific locations, including family plots cultivated with native plants like camas and later with non-native foods. Crossing the Cascades, they mingled with peoples of the Columbia Plateau, including the Okanagan and Yakama nations. Less fluid political borders registered slowly here. When the U.S. government negotiated the Treaty of Point Elliott in 1855 and ratified if four years later, the Sauk-Suiattle people disputed its terms. Many remained at home in the Cascade foothills at Sauk Prairie and the nearby drainages rather than relocating to the

Tulalip Indian Reservation thirty-five miles away. Tribal members wanted to take out homestead titles in their traditional territory, near present-day Darrington, Washington, rather than settle at Tulalip, a desire inconsistent with both federal Indian policy and the intent of public land policy.[7]

That homesteading laws intruded on Sauk-Suiattle country reflected growing economic and political incursions into the North Cascades. The Homestead Act of 1862 helped put the public domain into individual hands, a privatizing of the federal landscape seen as essential for improving and stabilizing the republic.[8] Reservation boundaries and homestead plots represented new lines on the land drawn by legal pens to create a new federal cartography. While independent Sauk-Suiattle families sought to legitimize their land claims in these mountain drainages, railroad surveyors, miners, and accompanying soldiers spread up through the mountains beginning with the 1858 Fraser River gold rush in Canada and accelerating after 1880 when miners in a second gold rush destabilized the Natives' homelands more. The best lands for settlement, including those in the greater Sauk River watershed, were already settled, a reminder that "settlement" really was a colonial resettlement.[9] Still, the mountains' ruggedness slowed the displacement of Native peoples.

Promoting private property was the business of government in the nineteenth century, whether supporting farmers or miners.[10] Turning the public domain into private domains had guided the republic's land policies since the Northwest Ordinance of 1785, before the states had ratified the U.S. Constitution. By the mid-nineteenth century, the American West's geography had stymied the progressive, linear story of homesteads inevitably replacing either wilderness or Native homelands. Walls of mountains and expanses of deserts complicated the presumption that the land would all be turned into small farms, while the isolation of these regions slowed their integration into the market.[11] Yet mountains contained mineral resources, and promises of wealth brought prospectors even to isolated ridges and valleys, like those in the North Cascades. Federal legislators encouraged this migration with the 1872 General Mining Law. Despite the passage of more than fourteen decades, this law continues to a large extent to serve as mining's legislative lodestar. It opened all of the public domain to mining, stating that "all valuable mineral deposits in lands belonging to the United States, both surveyed and unsurveyed, are hereby declared to be *free and open* to exploration and purchase." It allowed anyone to claim public land for prospecting; no payment for the minerals once found was required. At the time of the claim, prospectors needed to testify that they had invested $500 in labor and improvements, after which miners needed to invest $100 per year in labor or improvements to maintain the claim. The land itself could be patented—that is, formally deeded from the federal government—for merely five dollars per acre. (Miners could also work unpatented claims and extract mineral wealth but never acquire actual title.) It was a bargain.[12]

The easy terms, along with dreams of striking rich veins of ore, drew prospectors into the North Cascades, but results were mixed. By 1900 most mineral deposits had been discovered and claimed, and prospectors waded in mountain streams and trudged up steep ridges for minimal reward. Rushes flashed at places like Ruby Creek and Horseshoe Basin, bringing thousands of miners and hundreds of claims but only modest wealth.[13] Occasionally, such as at Monte Cristo, southwest of Glacier Peak, lead-silver ore paid out to Rockefeller family interests for more than a decade.[14] The other important exception was Holden. The history of this mine, about ten miles east of Image Lake just over the Cascade crest, stretched back to the Great Northern Railway survey in 1887, when Albert B. Rogers noted what appeared to be mineralized outcroppings. Rogers's report brought James H. Holden (with backing from Seattle's founding Denny family) to investigate the Railroad Creek drainage beginning in 1892. He finally found copper ore in 1896. Like those in other Cascades claims, Holden's minerals could not justify development right away. Inaccessibility, questionable profitability, and relatively low-grade ore delayed development.[15] The copper at Miners Ridge had been found by 1901 and claims made, but its promise would have to wait.[16] Despite generous federal laws, mining did not come to dominate many North Cascades landscapes. The exceptions seemed to prove the rule: this was timber, not mining, country.

Different federal measures drew other lines on the map. Besides the General Mining Law, another policy from 1872 changed priorities that allowed a counter-system to unfold—a public property system. In 1871 government-sponsored explorers in the Hayden Survey visited the land that would become Yellowstone National Park and feared that a railroad company or enterprising individuals might acquire this unique and wondrous landscape and close off public access and commercialize it. Beyond-belief descriptions of geysers and canyons inspired Congress in March 1872 to put 2.2 million acres into the world's first national park, establishing a precedent that some lands would remain in public ownership "for the benefit and enjoyment of the people." Other national parks would follow, and they tended to be places of magnificent scenery with boundaries drawn to exclude potentially valuable resources. Mount Rainier National Park, established in 1899 not far south of Glacier Peak, for example, centered on the mountain and its glaciers, not its heavily timbered foothills. Most Americans did not object to drawing park boundaries around beautiful scenic wonders where traditional economic activities would be hard to pursue.[17]

The most important federal program for the North Cascades by the turn of the twentieth century was the national forest system. In 1891 Congress gave the president the power to set aside lands from settlement to be maintained as forest reserves. Reformers had grown alarmed in the second half of the nineteenth century at the condition of nation's forests and especially at the devastation timber companies had left behind on cutover lands in the Upper Midwest.

To stave off a timber famine and to protect watersheds for both urban and agricultural uses, scientists and politicians generated momentum for forest conservation. Once empowered, presidents beginning with Benjamin Harrison wasted no time and set about reserving forests in western mountain ranges.[18] Washington's first forest reserve, the Pacific Forest Reserve, was established in 1893, centered near Mount Rainier, and included almost a million acres. It was a modest start. When the National Academy of Sciences sent a National Forest Commission to the West to investigate forest reserves and conservation needs, it recommended more reserves. President Grover Cleveland responded by creating several in 1897, including the Washington Forest Reserve of 3,594,240 acres (an area slightly larger than Connecticut) in the North Cascades. This was part of the notorious "Washington Birthday Reserves" where Cleveland withdrew more than twenty-one million acres reserved from the public domain on February 22, 1897, just before leaving office. Westerners' anger erupted, and, as William McKinley replaced Cleveland in the White House, Congress stepped in with the Organic Act of 1897 that suspended the reserves long enough to give individuals and corporations a chance to file legitimate claims. The law also identified three purposes for future reserves: improving forests, protecting water, and supplying timber.[19] Soon government agents studied the Washington forests, attempting to bring science to a relatively unknown region. The results were mixed. The initial U.S. Geological Survey report confirmed that although there was abundant timber in the reserved land, no market existed for it, and transportation links were all but absent. The mountainous terrain and "rapid and rocky" streams largely protected the North Cascades from timber exploitation in much the same way as the geography repelled mining development.[20]

Although Congress finally provided some instructions and limits on the new forest reserves, other reforms further changed their administration. In 1905 Congress transferred the reserves from the Department of the Interior to the Department of Agriculture, created the U.S. Forest Service, and placed Gifford Pinchot in charge as chief forester. Two years later the terminology was changed, and "forest reserves" became "national forests."[21] The secretary of agriculture, James Wilson, instructed Pinchot (in a letter probably penned by Pinchot himself) that "all land is to be devoted to its most productive use for the permanent good of the whole people; and not for the temporary benefit of individuals or companies."[22] This political rhetoric—a call to pursue the public good—set a clear direction that would be adjusted and stymied and redefined as history unfolded.

Like two people stumbling in a dark room, national forest conservation and national park conservation coexisted uneasily at first, as each worked out evolving ideas and as the public expressed its own ambiguous desires. The National Park Service (NPS) was created in 1916, forty-four years after Congress created the first national park, twenty-five years after the first forest reserve, and eleven years after the Forest Service started. In creating the first national park, Yellowstone, Congress instructed that the landscape was to be "dedicated and set apart as a public park or pleasuring-ground for the benefit and enjoyment of the people." By contrast, the USFS was guided by a utilitarian philosophy, intended to serve the "greatest good for the greatest number for the longest time," in Pinchot's words, evolving from the Progressive Era conviction that scientific experts could manage resources expertly to achieve long-term security. This orientation came from a fear about industrialization's costs.[23]

The NPS also grew out of such a fear. A growing number of urban Americans argued that individual and national character derived from the nation's frontier past. They justified wilderness preservation in escapist terms. American culture—at least its mythic white elements—required alternatives to the hubbub of the increasingly crowded, unhealthily polluted, and seemingly artificial nature of cities. National parks and wilderness would rejuvenate the crowds.[24] As the nineteenth century bumped into the twentieth, an organized movement emerged to celebrate and eventually protect wilderness. Outings clubs emerged—the Appalachian Mountain Club in 1876, the Sierra Club in 1892, the Mazamas in 1894, the Mountaineers in 1906—that took outdoor enthusiasts to the mountains, the first step toward building strong constituencies that would organize politically to protect access to wild mountains.[25] John Muir presided over the Sierra Club and symbolized its mission. A frequent essayist in the nation's leading journals and author of several books, Muir churned out pithy phrases about the rejuvenating power of wilderness, such as, "In God's wildness lies the hope of the world—the great fresh, unblighted, unredeemed wilderness. The galling harness of civilization drops off, and the wounds heal ere we are aware."[26] Muir helped cement the dichotomy between a healing wilderness and wounding civilization, between fresh nature and moribund cities. Muir died in 1914 at the end of a losing campaign to keep a dam from his beloved Yosemite National Park. Within two years, Congress created the National Park Service, its organic act paradoxically promising "to conserve the scenery and the natural and historic objects and the wild life therein and to provide for the enjoyment of the same in such manner and by such means as will leave them unimpaired for the enjoyment of future generations."[27] How to both conserve unimpaired and promote enjoyment has bedeviled the Park Service for more than a century.

So, while the Forest Service aimed to protect trees, the Park Service targeted views. At times, these missions overlapped, but they did not inherently do so, with disjunction evident in emerging agency cultures. Historian Hal Rothman characterized the USFS as bridging the individualist nineteenth century and the regulatory twentieth century guided by scientific efficiency, while the NPS evolved squarely in a twentieth-century culture of consumerism, where it sought to create and meet public demands.[28] Early Park Service directors aggressively poached scenic treasures from its rival bureaucracy since the Forest Service controlled millions of acres of the West's highest mountains and best scenery, a practice that rankled the Forest Service. The Park Service also knew that its survival depended on building a constituency of tourists, so it accommodated easy access, at first with railroads and then by catering to automobiles. The NPS pioneered its embrace of motor vehicles at Mount Rainier National Park, a mere seventy-five miles from Glacier Peak. The Park Service's mission was to preserve and promote, and it saw cars as doing both and so planned accordingly—structuring the scenery for drivers to see easily and send them on their way with good memories. By the 1920s, Henry Ford's Model T revolution, affordable cars, and ubiquitous roads combined with promotional campaigns that told the public to "see America first," put many thousands of Americans on the road to see wondrous vistas along a circuitous trail of grand national parks and monuments.[29]

The success of this culture of consuming scenery spawned critics, and many of the most prominent were in the Forest Service. Several foresters voiced early concerns about recreational development. Proliferating roads and cabins eroded the rugged backcountry's isolation, allowing elements of modernity to intrude.[30] So, while many conservation conflicts, then and now, pitted some sort of extractive use against some sort of preservation, the interwar period found contests over recreation priorities animating Forest Service personnel. Arthur Carhart and Aldo Leopold began thinking about wilderness in the late 1910s and early 1920s. Carhart, a USFS landscape architect, recognized that summer home developments planned for Trapper's Lake, Colorado, would mar both scenery and lake access, so he designated a buffer zone. This choice, while not preserving a wilderness ecosystem on a grand scale, gave the Forest Service its first taste of restraint in recreational development. Carhart stuck out in Forest Service culture and eventually left the agency, tired of lacking funding.[31]

Meanwhile, Leopold engaged intellectually with wilderness. In 1921 he famously defined wilderness as "a continuous stretch of country preserved in its natural state, open to lawful hunting and fishing, big enough to absorb a two weeks' pack trip, and kept devoid of roads, artificial trails, cottages, or other works of man."[32] He continued to refine his ideas, convinced that recreation formed a legitimate and sometimes the highest use of national forests. Leopold proposed protecting part of the Gila National Forest in New Mexico in wilderness conditions, and by 1924 the local Forest Service had designated the Gila Wilderness Area. By his reasoning, wilderness needed to be large and limited to "primitive means of subsistence and travel."[33] Specifically, Leopold contrasted national forest recreation with national park tourism, the latter being networked by roads. The national forests, in his view, fulfilled a different need, supporting that minority who wanted to avoid "all the automobile roads, summer hotels, graded trails, and other modern conveniences."[34] Although he did not remain in the Forest Service long, Leopold nudged its leaders to think about wilderness policy as part of the agency's general approach to managing land and as a counter to rising consumerism.

A third USFS employee, Robert Marshall, took wilderness and recreation further, formulating policy and playing a specific role in the North Cascades. Raised in New York, trained as a forester, and known as a radical, Marshall saw wilderness as a place for freedom, a site to develop masculine independence, and a political cause worth protecting.[35] Marshall earned a PhD in plant physiology from Johns Hopkins University and started with the Forest Service in a summer job in 1924 at Washington State's Wind River Experiment Station. He later held various positions in the agency and with the Bureau of Indian Affairs until his death in 1939.[36] Although his scientific training focused on technical silvicultural issues, Marshall developed a recreation specialty within the USFS. Tapped to contribute the recreation portions of *A National Plan for American Forestry,* known widely as the Copeland Report, Marshall articulated specific types of recreation opportunities and advocated for particular landscape zones, including wilderness. For wilderness he suggested a minimum size of two hundred thousand acres.[37] He built on existing Forest Service policy, Regulation L-20 established in 1929, which protected so-called primitive areas that district rangers managed. Working as head of the agency's Division of Recreation and Lands, Marshall helped to craft the newer Regulation U-1 (and U-2 and U-3, depending on the size of the tract to be protected) that gave the Forest Service the power to create wilderness areas in national forests where roads, commercial timber harvests, and permanent buildings would be prohibited. In place by the fall of 1939, these regulations strengthened the level of protection by putting their power in the secretary of agriculture or agency chief (again, depending on the size) rather than local district rangers.[38] Thanks to Marshall's work, the "U-Regulations" upgraded wilderness protection, making them more secure than ever, although administrative wilderness always remained vulnerable to administrators.

Meanwhile, as the National Park Service developed its scenic resources and the Forest Service provided its alternative to "windshield wilderness," as one historian termed the NPS approach, wilderness advocates converged to create a private organization to advocate for wild places.[39] The Wilderness Society was conceived by Leopold, Marshall, and others along a road outside Knoxville, Tennessee, in 1934 and formally launched in 1935 to advocate for federal protection of large roadless areas.[40] Its founders came to this point from different pathways but were united in their fear of mass recreation—symbolized by the automobile—invading the undeveloped spaces left.

In the inaugural, September 1935 issue of the *Living Wilderness,* the organization's newsletter, just nine months after the organization incorporated, Marshall highlighted the North Cascades as a representative case. He promoted the region's scenery and worked to alert the public. In "Three Great Western

Wildernesses: What Must Be Done to Save Them?," Marshall presented an expansive vision for the North Cascades region and its "backbone," a place that protected the region's important scenic qualities the same way vertebrae protect the spinal cord. Marshall's wilderness encompassed the Cascade crest, which presented an opportunity "to preserve for wilderness travel one of the most stupenduously scenic areas in the United States." Marshall anticipated objections and countered them in advance. Hearing the common message that the majority of people wanted automobiles and roads, he acknowledged that wilderness would deprive motorists from seeing this place. However, he thought it "even more important that no unfair monopoly of outstandingly beautiful Northern Cascade scenery be given to the motorist," and he pointed out a diversity of beautiful places in the region that remained accessible by road. Marshall ended his essay asking for readers to write to the Forest Service and the Bureau of Public Roads, encouraging them to set the land aside without roads and to write letters to policymakers and other influential people. Just so, the wilderness battle was enjoined, and the North Cascades was present at the creation.[41]

Important context for Marshall's proposal lurked in the background. First was the Park Service. During the New Deal, President Franklin Roosevelt's response to the Depression, Secretary of the Interior Harold L. Ickes moved to consolidate conservation within his department. One effort included expanding the Park Service. In Washington State, the NPS gained control of Mount Olympus National Monument when executive reorganization moved all national monuments to the Park Service. At the same time, federal recreation reports, issued by Ickes, described potential new national parks. Eventually the proposal emerged for Ice Peaks National Park, which incorporated the volcanic peaks in the Cascades: Mount Adams, Mount St. Helens, Mount Rainier, Glacier Peak, and Mount Baker. Except for Mount Rainier, which already was a national park, all this land would be taken from the Forest Service. Nothing could have been designed more purposefully to raise bureaucratic hackles.[42]

The Forest Service provided its own context. By 1931 the North Cascades contained a USFS-designated "primitive area" around the Picket Range, a recreation area centered on Mount Baker, and a recreation unit encompassing Glacier Peak. In all, this amounted to nearly half a million acres dedicated to one degree or another to recreation.[43] But these areas could be enlarged or eliminated. In his essay in *Living Wilderness,* Marshall praised Forest Service chief Ferdinand A. Silcox, who allowed the agency to expand the initial primitive area to include 801,000 acres in the North Cascades Primitive Area, a strip of wild land that touched the Canadian border.[44] But Marshall saw this designation as insufficient and wanted to incorporate a much larger area with much greater potential just south of the strip, an area that included Glacier Peak. Fred Cleator, the Forest Service recreation planner whose report opened this chapter, was responding directly to Marshall when he wrote in 1936 about Glacier Peak: "In recent years there has come a strong demand, principally from Wilderness enthusiasts, the country over, that the big country adjacent to the Cascade Divide from Glacier Peak north should be left as nearly as possible in its natural state."[45]

Cleator recognized that some conservationists were targeting the enthusiastic road-building that had become common in the preceding decade. The Depression-era relief workers in the Civilian Conservation Corps (CCC) built bridges, graded mountainsides, and paved roads seemingly everywhere. Roads took firefighters and camping families into the woods, but critics in and beyond the Forest Service doubted that the roads' benefits outweighed their costs.[46] Cleator's report needs to be understood in the broader critical context of commercialism and ease in the outdoors. Characterizing these perspectives, Cleator rhapsodized about those who favored wilderness recreation—"the increasing class of humanity who find the ordinary recreation too stifling and artificial, who wish to get away from motors and resorts, and who want things exactly as Nature built them or grew them"—in rather effusive rhetoric for a bureaucrat.[47]

Cleator praised Glacier Peak's suitability for wilderness designation and aimed for permanence. He described how the area varied "from the extremity of ruggedness to friendly mountain meadow types, containing now a natural biotic balance, which it seems should have excellent chances of perpetuation regardless of changing future administrations and policies." Cleator understood not only what made wilderness attractive but also the shortcomings of Forest Service policies that lacked permanency. Both the agency and its constituents, such as the timber industry, wanted certainty to aid long-term planning.[48] Cleator hoped that this area would remain wilderness because of its natural features and comparative lack of commercially valuable resources. Its timber and mineral resources appeared too small, with transportation costs too high, to outweigh the value of wilderness.[49]

However, Cleator recommended some developments. Small shelters and better trails would improve recreation. He believed some roads would be necessary. Although he wrote that "the spirit and intent of this plan [is] to permit the least possible amount of road building within the area," the needs for fire protection and getting ore "mine-to-market" meant that a few incursions seemed necessary. In particular, he highlighted the need for a road along the Suiattle River to serve the Miners Ridge holdings, an incursion that would continue to figure in this story.[50] Despite these exceptions, Cleator strongly favored wilderness. His conclusion, in full, was a remarkable statement that reflected a moment in the agency's history when vigorous wilderness protection mattered: "It is, and should be for all time, the function of every Forest Officer who is charged with any responsibility involving this area, to preserve the Wilderness Area ideals, to act promptly and effectively in safeguarding the area from vandalism, from public advertising, from improper commercialism or other malpractice, as well as to protect it from fire and pollution, and must do his best to help safeguard human life and limb in this rugged and dangerous country."[51] How to mine the ore from Miners Ridge without pollution remained unresolved and little remarked on. But without equivocating, Cleator voiced a strong sentiment for USFS wilderness.

In the national office, assistant USFS chief Leon F. Kneipp read the wilderness proposal with a favorable but conflicted eye. He had promoted recreation in the agency and directed an inventory of roadless lands beginning in 1926.[52] But, however much he agreed in principle with the wilderness report, concerns gnawed at Kneipp. The Suiattle River Road, he wrote, "would constitute a very definite and considerable invasion" but seemed necessary for fire prevention and mineral extraction. Cleator's report emphasized the improved view of Glacier Peak the road might offer, but Kneipp recognized "a certain inconsistency with the primitive-area idea to provide motor transportation to the key feature of the area."[53] This primitive sensibility in the Forest Service aligned with the Wilderness Society's ideas and demonstrated how roads transected most wilderness debates. But the presence of timber along the river forced Kneipp to wonder about the future. If the timber might one day be cut, the primitive area boundaries might need rethinking to avoid conflict with the road and putting into question "the desirability of formalizing the dedication in the manner proposed" by Cleator. Then again, Kneipp allowed that perhaps only "such formal dedication is the only feasible means of making permanent the plans and restrictions embodied in the report."[54] Evidence from within the agency shows a struggle over the fundamentals of primitive area principles and genuine questioning whether such a designation was appropriate or essential. Perhaps it was too much and the agency could not figure it out, because no resolution remains in the historical record. Instead there is just a rejection of Glacier Peak Wilderness Area as Cleator had envisioned.[55]

Bob Marshall stepped in literally by surveying the North Cascades himself. Marshall built a reputation as a prodigious hiker; an early coworker recalled that Marshall would sometimes hike more than forty miles on his days off, a behavior this colleague saw as fanatic.[56] Just six weeks before his death, in September 1939, Marshall traveled with Cleator through the North Cascades. When the party climbed from the Suiattle River up toward Miners Ridge, Marshall refused a horse, favoring instead the steady steps of his own two legs, reckoning that "the horse was still a mechanism of a sort." Cleator recalled that Marshall "knew as well as I that freedom to ride and look about, to get one's eyes off the trail, did not always compensate for the responsibility and grief one may experience with a horse." High up, the party enjoyed huckleberries, and Marshall savored solitude, withdrawing from Forest Service fire circle camaraderie "to commune with his idolized back country. . . . He seemed to feel a great need to use every available hour to absorb raw Nature." Along the way, a Forest Service employee kept trying to include Marshall in photographs to show human scale, but he refused: "We are privileged to be looking at a billion years of the Creator's work. Why should I, a mere human, have the temerity to compete with this masterpiece?" Marshall wrote Cleator afterward, characterizing Glacier Peak country as "marvelously beautiful" and confessing that he knew of no other "country that surpasses in beauty the Northern Cascades."[57] Shortly after this hike, Marshall fell ill and died, just shy of his thirty-ninth birthday.

While Marshall tramped with Cleator, the secretary of agriculture adopted the U-Regulations Marshall had created to designate wilderness in national forests. Places like the North Cascades possessed values more than the commodities they contained; this was Marshall's main point. In his inspection report submitted subsequently, he argued that "no part of the whole United States is so well adapted for a wilderness as the country between Stevens Pass and Hart Pass," a large area that encompassed Glacier Peak.[58] This whole big country, marked by peaks and passes as magnificent as anywhere, deserved wilderness protection. Knowing of plans then circulating, Marshall feared Park Service designs, telling Irving Clark, a stalwart Northwest conservationist, that Forest Service wilderness would be better: "I know and you know perfectly well that if this area should be made a park it would have roads extended to into its heart."[59] Conservationists concerned about North Cascades wildness would continue to question which agency offered the best options, the strongest protection, the greatest political strength. This narrative thread winds through this story from start to finish.

The summer before Marshall's last hike, the Wilderness Society poked at the Forest Service, no doubt egged on by Marshall. Robert Sterling Yard, one of the organization's other founders and its first president, wrote to USFS chief Silcox to inform him that the Wilderness Society had passed a resolution that urged the agency to create the nation's "biggest, and in some respects the greatest forest roadless area" by incorporating all the northern Cascade crest south to Stevens Pass.[60] This reiterated Marshall's plea in the first *Living Wilderness* three years before. So, when a road edged into this region, Yard was frustrated since Silcox had assured him the agency was studying the area for wilderness designation. From their earlier exchanges, Yard assumed the Wilderness Society would be provided the opportunity to share its ideas if the Forest Service contemplated changes to the North Cascades. Feeling betrayed, Yard slammed his cards on the table. Many society members, especially those rooted in the Northwest, felt "indignant," believing "the only safe way to preserve the wilderness is to put it into a national park." Yard played a strong hand designed to stir up trouble. The Forest Service and National Park Service, like two roosters in the henhouse, danced around each other for years, and Yard exploited the circumstances.[61]

To good effect. From the Washington, D.C., office, Silcox ordered the regional forester, in no uncertain terms, to start no new road projects in the region, characterizing the agency's actions as double-crossing the Wilderness Society. This was a remarkable sentiment. The Wilderness Society, barely four years old, had, it seems, moved the Forest Service chief by a simple letter. Silcox informed the regional forester that two rangers, one of whom was Bob Marshall, soon would be in the region to study the question of boundaries. Silcox hoped for some finality but was unwilling to rush. Then, near the end of 1939, Silcox died. The two strongest Forest Service advocates for a wilder North Cascades died before completing their legacies.[62]

Unrestrained, the regional office pressed forward. A series of stopgap designations protected the Glacier Peak area. A 233,600-acre Glacier Peak–Cascades Recreation Unit existed, although this designation merely announced that the USFS deemed recreation the highest use there. In 1940 the agency created a larger (352,000-acre) Glacier Peak Limited Area. "Limited area" was a label unique to Region 6. The region, which comprised Washington and Oregon national forests, used "limited area" in places considered off limits to development until the agency decided otherwise.[63] The agency referred to it as a "stop-look-listen" approach.[64] But Harvey Manning, a longtime Northwest conservationist known for his acerbic wit, once characterized "limited area" as meaning "we haven't yet figured out where to put the logging roads." Manning could turn a phrase for a cause, but he recognized the ephemerality of limited areas: they could be eliminated from the regional office any afternoon.[65]

As the Forest Service drew lines across the Cascades, the debate moved beyond the agency as Northwest conservationists gathered. In 1940 the conversation helped to clarify—and perhaps harden—the distinct desires and perspectives conservationists held for Glacier Peak and the future of the North Cascades. They overwhelmingly focused on recreational potential for Cascades wilderness, but occasionally someone referenced scientific value. Irving Clark once envisioned the high Cascades as a "research laboratory." A tall thin man who was active in liberal causes like civil liberties and a longtime active member of Seattle's alpine club, the Mountaineers, Clark had long fought against incursions by timber companies on the Olympic Peninsula. But now he stood strong for North Cascades wilderness. Clark wrote to a U.S. representative from Washington State, "The region of the North Cascades should now be permanently dedicated to its highest and best use,—its scientific, educational and recreational use as a wilderness."[66] So sure was Clark in his convictions that he advocated ignoring local interests and miners when he urged the Forest Service to use its property rights without consulting others. Shortchanging democratic input like this was necessary in Clark's mind to counter the prevailing local economic powers. At the same time, growing tired of the Mountaineers' drift away from politics, Clark asserted stronger political organizing marshaled through the Wilderness Society, where he sat on the governing council thanks to his friendship with Marshall.[67]

Clark became the strongest local force for Glacier Peak wilderness at this time, a few months before Franklin Roosevelt won his unprecedented third term as president. He used his energy to press the Forest Service's leading recreation specialists in the Region 6 office and the national office, F. V. Horton and John Sieker respectively. Both Horton and Sieker, persuaded by local commercial interests, envisioned roads as inevitable. A future road, or several, crossing the North Cascades, along with mine-to-market roads winding down from mountain ridges through river valleys, would preclude the giant wilderness Marshall and Silcox wanted and for which Clark advocated. Although wilderness advocates disagreed, Forest Service personnel foresaw plenty of wilderness left over after these road projects, some four hundred thousand acres in the Glacier Peak area alone. Some in the agency, such as Cleator, might regret the incursions, but they generally thought that some wilderness was better than none. In truth, the Forest Service was caught. Directed to support local economic interests, the agency followed a policy of segregating mineralized regions to avoid conflict with miners, and public land law favored miners. The Forest Service saw no other options.[68]

Rumors swirled and confused and conflated schemes and interests. For instance, many in the public thought the Forest Service's interest in Glacier Peak was a ruse the secretaries of agriculture and interior had devised to create a North Cascades National Park, a possibility that assumed a shared vision while ignoring the deep rivalries between them. The press and chambers of commerce lightly threatened Horton, saying they would stir up opposition if protected areas prevented access to favored local passes. Horton wished to avoid hassles for the secretary of agriculture but anticipated a "dog-fight between the wilderness people, the mining people, and the recreationists."[69]

This triad of interests is worth noting for two reasons. First, it shows how "wilderness people" were not synonymous with recreationists. That is, different types of recreation needed to be accommodated. Some recreationists wanted roads to mountain passes to make for easier camping or hunting, a prospect anathema to the entire wilderness project. Second, it neglects timber interests. Northwesterners would become accustomed to seeing the timber industry as the main wilderness opponent. But in this era before World War II, at least in the North Cascades, at least for now, commercial timber operations merited little attention. Although they would appear soon, it is important not to read a dominant presence backward through time. When Glacier Peak wilderness first became a public issue, mining more than logging concerned wilderness proponents.[70]

This hierarchy of use was confirmed with an important exchange in 1940 between the Wilderness Society's president, Robert Sterling Yard, and Forest Service officials. C. M. Granger, who was then serving as acting chief, recognized earlier hopes: "I know it was Bob Marshall's wish that the entire Glacier Peak area be made into a wilderness, and I would have liked to see that accomplished had it not been for the incontestable mineral values involved, but, as you know, our general rule is not to include in wilderness areas lands of high mineral value."[71] Yard informed Granger that the society rejected the Forest Service removing protection on corridors in the proposed Glacier Peak wilderness. Yard leveraged Marshall's and Silcox's recent deaths and asked what could justify denying Marshall's last request.[72] Yard posed a series of questions about mining there that he acknowledged might cause the agency "a great deal of trouble," but he told Granger that he hoped they would sort out the relationship between the agency and the Wilderness Society, saying the two bodies "believe in the same ideals."[73] This final point might be read simply as an effort at finding or creating or pretending they shared common ground, but it demonstrates how recently the Wilderness Society had been founded (only five years before) and Forest Service U-Regulations adopted (the preceding year). Each group was still negotiating how to understand—and influence—the other. Glacier Peak wilderness helped advance the relationship.

The Forest Service coordinated between the national and regional offices to answer Yard.[74] The Wilderness Society wondered if chambers of commerce or other economic groups pressured the agency. The answers were revealing. The regional forester Lyle Watts denied that any individual or corporation seeking to mine near Glacier Peak had influenced his recommendations. Yet, two paragraphs later, he explained how people from nearby cities, both east and west of the Cascade crest, "definitely state that there is a need for roads" to get into the high country.

> Their argument, which appears to us a logical one, is that if the Glacier Peak area is to be enjoyed it must be made more accessible by having roads to the edge of it through the uninteresting fringe which is now a barrier except to the few who are extraordinarily equipped physically or who are wealthy enough to hire pack outfits. A sixteen-mile hike up hill, for instance, from the west, is a pretty effective barrier to all but a few people. It creates a hierarchy based on wealth or physical energy. Even with the exclusions there remains sufficient area to test the stamina of those who have the physical qualifications to challenge the real wilderness without benefit of an elaborate safari.[75]

This passage's threads are worth unraveling. First, Watts indicated that the chamber of commerce types Yard worried about were precisely those who advocated the road to Indian Pass and Buck Creek Pass.[76] Second, he assumed that wilderness largely was about proving one's physical stamina—that "real wilderness" constituted physical tests, which did represent a significant component, but by no means the only one, of wilderness advocacy.[77] Third, Watts referenced the cost barriers to wilderness travel. The supposed elitism of wilderness advocates often centered on this charge, so it is worth considering. To take time off work for the days needed to get deep in the backcountry did require a certain degree of wealth. Requirements such as gear and food made it difficult to spend much time in the wilderness without a packtrain, a claim that reminds us of the revolution in gear that allows today's backpackers much greater freedom.[78] Also, though, many people of this era managed mountain time fine without elaborate packtrains. So this elitist charge always was more complicated than those deploying it acknowledged. Last, in following the chamber of commerce logic, Watts suggested that a low-country forest was merely an "uninteresting fringe" and "barrier," a perspective profoundly at odds with Northwest wilderness proponents who valued the woods as much as the high alpine meadows and ice-covered peaks.[79] The Forest Service layered wilderness with support and criticism, and within those perspectives were simplified complexities.

Agency leaders used this opportunity to explore "fundamental problems" related to wilderness and forest administration, anticipating many future instances when the agency might face off with the Wilderness Society. In the main, Granger reiterated all of Watts's points except for the desire of locals for a road up into the high country. Granger seemed to recognize better than Watts that Yard would find this local desire for easier access the very problem the Wilderness Society fretted about and opposed. However, Granger contradicted himself by arguing first that the USFS faced "no direct pressure . . . from either individuals or corporations" but then indicating that the Miners and Prospectors Association opposed any restrictions to their activities. Granger also turned to what he called "general principles." Granger reminded Yard of the Forest Service's achievements, citing its seventy-four units covering fourteen million acres designated in primitive, wild, or wilderness categories. When deciding what areas merited inclusion in the agency's wilderness, foresters applied "sound principles of land use planning," Granger told Yard. Jabbing gently, Granger said the agency always considered the general public and "special interest groups, of which you recognize, of course, the Wilderness Society is one." But the real issue on the ground level here was that mining laws carried special power. In short, they superseded all forest laws; the Forest Service could not stop mining exploration and development if it wanted to. Excluding known mineralized areas from wilderness made for good practical management so the agency would be saved from "having frequent eliminations made from established areas" when mines developed. The Forest Service knew such exclusions would irritate Yard, the Wilderness Society, and similar constituents. But then Granger reminded Yard that the agency had never approved wilderness at Glacier Peak; they were merely holding it in status quo until further study, serving only to encourage wilderness advocates to press the agency to designate it appropriately. Despite the mixed message he delivered, Granger hoped the Wilderness Society might soon develop confidence in USFS wilderness policy.[80] Developing that confidence would take time, and soon a test presented itself.

World War II arrived with its weighty issues and influenced the Forest Service's and conservationists' hierarchy of priorities. It would not be the last time international relations would affect this site in eastern Snohomish County. Although prospectors had located copper as early as 1901, the costs to develop a working mine, compared with the potential profit, thwarted development. By the late 1930s, though, the M. A. Hanna Company was actively exploring at Miners Ridge, calling its operation the Glacier Peak Mine even though it remained only an exploration site. Hanna's tests revealed great potential. In 1942, a representative from the U.S. Bureau of Mines estimated Miners Ridge containing sixteen million tons of ore, including a bit more than 1 percent copper, along with smaller amounts of molybdenum and scattered silver and gold. Seventeen thousand feet of drill holes and six hundred feet of tunnels showed the results of the four previous summers of diamond drilling. According to the Bureau of Mines, investing $5.5 million would be sufficient to develop a commercially viable mine and build the five-thousand-ton-per-day flotation plant needed on the site to process the copper ore. Experts anticipated profits within three years of full-fledged mining. Thus, the Bureau of Mines formally requested that the Forest Service approve a road up the Suiattle River to launch the project. Although the regional forester, facing "heavy pressure," immediately recommended the new road to the agency's acting chief, the Washington office tapped the brakes.[81]

By 1942 Earle Clapp was serving as the USFS acting chief. He had earlier headed the agency's research division, which he expanded and rescued from irrelevance.[82] Nothing in Clapp's record suggested any special interest in wilderness. Rather than rubber-stamping the regional forester's recommendation, Clapp told the Bureau of Mines director that approving the road presented the Forest Service with a problem. The road would cut into an area "of such superlative natural and scenic interest and recreational opportunity" that it had been considered as wilderness, a designation that disallowed roads and vehicles. The agency had promised wilderness groups, Clapp explained, that they would be consulted before any changes were implemented, a commitment carried on from Silcox. Clapp revealed the importance of cooperating—at least rhetorically—with groups such as the Wilderness Society.[83]

Clapp recognized that "conditions have been so changed by the war that pre-war commitments . . . cannot be meticulously observed in ways detrimental to national welfare and security." Still, the acting chief was not ready to fire up the diesel engines on the bulldozers, carve out the forests, and scrape a road into alpine meadows near Glacier Peak. Clapp demanded assurances that the minerals were necessary and could be developed to aid the war effort, which would "transcend all" other needs since such actions would cause "irreparable impairment of one of the few remaining wild areas." Could the Bureau of Mines provide such assurance? Clapp's attention to the irreparable nature of roads and his attention to promises made to conservationists indicated that even in a wartime emergency wilderness values might receive a fair hearing.[84]

But perhaps not that fair a hearing. Assurances came cheaply and with little evidence other than the word of corporate presidents and government bureaucrats. The War Production Board—created to coordinate resources and production to meet wartime needs—assured Forest Service administrators of the mine's necessity. In late September 1942, the Bureau of Mines hoped that approval might come in time to move machinery into the mountains before snow arrived.[85] It took only two letters from the Bureau of Mines for the Forest Service to approve the road.[86]

In the meantime, Forest Service officials consulted with Yard of the Wilderness Society. Perhaps this served only as a courtesy, but the record reads as genuine consultation. In fact, the Forest Service in its own correspondence characterized Yard as acquiescing, suggesting that he had some sort of power to surrender.[87] Various Forest Service personnel at least paid lip service to wilderness concerns. For instance, in writing to Yard to let the Wilderness Society know the road would be built, Granger empathized: "You can imagine the reluctance with which we have entertained the proposal to bisect the potential Glacier Peak wilderness area with a road. But in the circumstances created by the war, opposition would be difficult to justify. As I recall it, that was approximately your own conclusion."[88] Then, when regional forester Lyle Watts, who soon would be elevated to chief forester and who had been prepared to approve the road immediately, informed Irving Clark of the agency's decision, he described the road as invading the area under study for wilderness.[89] Such language—Watts's exact word was "invades"—demonstrates marked ambivalence in the Forest Service.

The road through the Suiattle River corridor and up to Miners Ridge then seemed a foregone conclusion. Yard expressed his regrets.[90] Clark showed impatience and frustration in periodic letters to the Forest Service inquiring about the project's progress.[91] But then came a reprieve. The War Production Board suspended the project in February 1943 before new road construction began. Extant documents do not explain the reversal. By war's end, local foresters reported that the project was "now abandoned and is not likely to be reconsidered." Not only that, the regional forester in 1945, H. J. Andrews, proudly told Clark that Washington's and Oregon's wild areas came through the war "in good shape," without any "unauthorized developments." Region 6 even created two new wild areas during the war.[92]

Although the Wilderness Society and the Mountaineers had not mounted a fierce protest during World War II to stop the mine from getting established near Glacier Peak, the organizations and their members surely breathed easier having withstood the threat. As the war concluded, the Forest Service contemplated priorities not related to wartime production, including reconsidering Glacier Peak as wilderness. Regional forester Andrews assured Clark that the Forest Service would keep conservationists informed.[93] Conciliation seemed the order of the day. But the war had tested everyone's patience, and the postwar era would sharpen the edges of the relationships. The first test concluded, and Miners Ridge seemed safe enough.

In the 1940s the North Cascades existed in a relatively novel condition. The region's physical geography had remained basically unchanged since the colonial powers had arrived in the late eighteenth century. But claims by new empires and cessions by old Indigenous ones, followed by corporate and governmental enterprises, had combined to remake the Cascades. By the twentieth century, new institutions overlay this landscape in the nation's northwestern corner. The federal government's conservation efforts comprised competing factions and evolving ideas as agencies developed their missions and responded to both congressional mandates and a public that voiced its own shifting priorities, sometimes finding common ground. Abstract principles, though, applied on the ground in real places like Miners Ridge. The area around Glacier Peak became a site of conflict and compromise. The U.S. Forest Service controlled it; the National Park Service envied it; mining companies claimed parts of it; and outdoor recreation enthusiasts wanted it for themselves—some with roads to acquire the view easily, some wishing for primitive conditions. The minerals lodged in rocks and the beauty lodged in minds meant that competing ideas and purposes clashed here. World War II demonstrated how quickly things could change and change again based on factors away from the mountains, foreshadowing the history to come.

After fending off the Park Service and surviving the war, the Forest Service was poised to extend and strengthen its protection of Glacier Peak. But a powerful set of citizens pulled on the reins that the agency had long considered its own. For a decade starting in the mid-1950s, local and national wilderness proponents made Glacier Peak a case study of citizen activism, creating a powerful narrative that guided long-term debates and management priorities.

TWO | The Wilderness

The inner world of mountains has remained almost unknown and unspoiled. Its sounds remain solely those of wind and moving water. The fact is the product of accident, no credit being owed to the policy of men, but the change is good fortune, for the region is among the superlative areas of the continent.

Grant McConnell, "The Cascade Wilderness" (1956)

AT FIRST IT SEEMS an unlikely pairing: a university professor and a remote Northwest wilderness area. But Grant McConnell and his wife Jane purchased a small plot in Stehekin Valley in 1945 just as World War II ended, an act he later called his "ultimate commitment to the region that has been almost central to my life."[1] McConnell, a political scientist then at University of California, Berkeley, but soon to leave for the University of Chicago, brought a prescient and insightful perspective to the North Cascades and understood what was at stake there. He and Jane gained firsthand regional knowledge from living at Stehekin and tramping through the nearby ridges and isolated valleys, experiences that gave McConnell much to say. Writing in the *Sierra Club Bulletin* in 1956, McConnell sought to share what he knew and believed with a sympathetic audience.

In "The Cascades Wilderness," beautifully illustrated with eight Philip Hyde photographs, McConnell portrayed a striking region. The North Cascades could be described only in superlatives: "the nation's finest alpine area," "one of its most untouched primeval regions," "a sanctuary, one of the country's last and perhaps its greatest"—all in the opening paragraph. Their characteristic "high peaks, deep valleys and rushing water" meant that economic development did not penetrate far into these mountains, a historical accident relished by McConnell. He recalled the history of efforts to establish a commercial foothold there, as prospectors with illusory dreams had crawled over the mountains searching for easy wealth. In 1956 the timber industry eyed the woods too, though McConnell reckoned the timber was poor quality and of "limited value." Pressure also mounted to build roads and resorts. McConnell sensed that this was all toward only "transitory gain."[2] This pressure to develop the land would enact a tragedy here.

McConnell understood the North Cascades not as a source of economic sustenance but as a spiritual reservoir. Like others before him, he used the word "sanctuary" deliberately. Besides the mountains' beauty and abundance, they represented something more, something beyond human values, something holistic. McConnell saw the North Cascades as "the true climax" of the entire mountain range and claimed the terrain as "a unity," a place whose value came from the "integral parts of a magnificent whole." Here was a place that by force of accident remained "unexplored" and thus represented opportunity. The key question centered on what that opportunity entailed. Was it the opportunity for people to explore "the haunting quality of the region?" Or was it "easy entry and exploitation" for "economic development and the demands of politically minded entrepreneurs?"[3] Answering the question was easy for McConnell, and he represented a growing group of postwar Americans who drew on ideas from Bob Marshall and others who valued wilderness.

In the final paragraphs of his essay, McConnell shifted from descriptive natural and human history to an assessment of the Forest Service. The shift was natural both because of his personal concern for the North Cascades and his academic training. Two years before, he had published a major article on conservation where he characterized it as having two streams:

> The one tends to emphasize an accommodation of uses, each of which may be presumed to stand on more or less even footing, except as one or another use appeals to greater or smaller numbers of people. Given the character of American society, this amounts to an emphasis on material values. The other elevates certain values to a preferred position, no matter how few the people to whom these appeal. It elevates nonmaterial above material values as a matter of principle.[4]

The Forest Service, McConnell believed, faced a major test over how this split conservation would play out in the North Cascades. Much of the Glacier Peak Limited Area awaited reclassification, potentially as a USFS wilderness area with stronger—although not foolproof—protections. But McConnell worried. The boundaries were poorly drawn and conceived, not following topography, excluding prime areas, and otherwise reducing protection. These decisions sacrificed the whole, which McConnell—like Marshall before him—recognized as the real and unique benefit of this place. McConnell understood the agency's multiple-use policy as "one of the genuine glories in the management of natural resources set forth by any nation," and he praised the Forest Service's goodwill and its rangers' "spirit of dedication." But times changed, and policies could be recast in light of new circumstances. Multiple use had to "be made relevant to the needs of today," McConnell wrote, as the human population grew and values evolved to favor "undesecrated remnants of our natural scene." McConnell wanted a multiple-use policy that recognized "that all values cannot be mixed without the extinction of some by others." Instead policy ought to deliberately zone areas for "their highest purpose," which could be scenery. McConnell wrote, "A region as splendid as any in the nation, one unique in alpine character and beauty, has by accident been preserved as real wilderness. It remains to be seen whether the intelligence of man can do as well through policy."[5] That view guided many activists in the coming years.

McConnell's *Sierra Club Bulletin* article coincided with a pivot in the campaign to protect Cascades wilderness. The National Park Service had expressed occasional interest here, but the Forest Service had controlled most of the area for a half a century, displaying strong, if piecemeal, protection of the northern mountain scenery through various "primitive," "wild," and "limited area" designations. Yet in the 1950s much was changing in the agency, the public, and the relationship between the two. More and more people like McConnell spent time climbing mountains and hiking ridges in a postwar recreation boom.[6] They hoped these peaks and valleys and rivers would stay wild so their grandchildren could enjoy them. The Forest Service faced pressure from this public, as well as from the timber industry that looked more and more to public forests for sawlogs.[7] These tracks converged, or crashed, with heightened controversies and consequences. By confronting the Forest Service here, public conservationists like McConnell affected policy decisions that shaped the land. The political context of Glacier Peak, then, transitioned in the 1950s and into the early 1960s.

Calls like McConnell's generated a public narrative about wilderness that attracted many passionate adherents whose activism shaped and constrained Forest Service actions. During the decade that followed McConnell's *Sierra Club Bulletin* article, a network of wilderness activists grew, with a narrative of action that would eventually carry northwesterners into the fight against Kennecott Copper. But first activists had to confront the Forest Service. During this period, the North Cascades became nationally known, which aided later campaigns. It was a time when federal agencies and local activists negotiated through evolving policies that reflected citizens' new values and new political power.

Conservationists were moving from defense to offense, but they still operated mostly in an arena where the Forest Service held power over wilderness preservation.[8] The early ideas from agency employees like Aldo Leopold, Arthur Carhart, and Bob Marshall, along with the policies being implemented, gave the Forest Service a wilderness legacy more formal and developed than that of the National Park Service. By the 1950s, though, changes in Forest Service priorities, new economic demands, and burgeoning conservation advocacy meant an increasingly tense relationship between the federal government's best—albeit imperfect—wilderness champion and the activists who aimed to secure an indelible place for wilderness on public lands. The problem of how to fit both wilderness and recreation into the Forest Service's "greatest good" orientation stymied the agency and eventually frayed the ties between agency administrators and conservationists. When the Forest Service redrew wilderness boundaries in the North Cascades, the lack of clarity and the inconsistencies in policy became as obvious as Glacier Peak's summit on a summer's day.

The Forest Service's U-Regulations, crafted by Marshall and adopted at his urging in 1939, required the agency to reclassify national forest roadless areas across the country, including in the Pacific Northwest. The agency protected fourteen million acres from roads and logging with its L-20 regulations in effect from 1929 to 1939; the U-Regulations required it to reclassify those sections and hold hearings to gain public input. In Region 6, the first major postwar reclassification was contested near the Three Sisters peaks in the Oregon Cascades, where the Forest Service opened fifty-three thousand acres of wilderness to timber harvesting. The reclassification process exposed the ways that U-Regulations failed to secure strong wilderness protection; administrators could—and did—change boundaries and open once-protected forests to timber sales. From 1951 to 1957, wilderness advocates from Oregon and beyond, led by Ruth and Karl Onthank, worked to stop the USFS from opening further lands to logging. They failed, and conservationists learned firsthand the power the bureaucracy wielded in administering wilderness boundaries. The specter of that power to take away protection remained with conservationists as they confronted the Forest Service in the North Cascades while reclassifying Glacier Peak.[9]

The agency as a whole struggled to place wilderness squarely within its policy purview, yet agency employees were not entirely insensitive to wilderness. A forest supervisor in 1954 allowed, "The idea of setting aside and preserving some of the unspoiled areas in the country to be enjoyed in their natural state by future generations certainly has merit." The more he wrote, the more he warmed to the idea: "Most of us in the Forest Service as individuals are sympathetic to the Wilderness Area concept." Turning to Glacier Peak specifically, he confessed that it "of course has long been recognized as a strikingly beautiful remote area with wilderness characteristics that qualify it for dedication as an excellent Wilderness Area." Still, the forester reflected the overriding ethos of the Forest Service, which had long sought to support what it determined to be "the needs of the public *as a whole.*" He acknowledged, "It is certainly right that representative tracts of virgin timber be preserved for all time as part of Wilderness Areas. However, timber and other values of the national forests are very important to local economies and to the national economy." Predictably, he relied on the agency's guiding aphorism, masked as unambiguous policy, saying, "such issues are decided on the basis of the greatest good to the greatest number in the long run." In other words, wilderness is fine, rangers like it too, but national forests serve economic development first and last.[10]

But, like beauty, the greatest good was in the eye of the beholder, and as local foresters studied the wilderness question around Glacier Peak in the 1950s, they confronted real places and dilemmas, not abstract ideas. In 1950 the Mt. Baker National Forest planned for changes, and wilderness values appeared in the details. A district ranger rejected a proposal to move a boundary from a river to the ridge above, a shift that would have opened "a large stand of old growth timber." According to the ranger, this "impressive but limited" timber stand needed to be protected "in its natural state for the benefit of those who appreciate natural beauty." Logging here would result in "too great a cost in the loss of natural beauty." The forest supervisor concurred. Sometimes beauty mattered.[11] But generally when the Forest Service recognized recreation—stating the likes of "public sentiment—highly favorable toward recreation" or "it is believed that public sentiment is in favor of setting aside this area for recreation purposes"—this meant increasing access, which meant extending the road system and ignoring the growing number of conservationists who found the best recreation in places without roads.[12] Struggles over these places and roads mounted through the 1950s and crept closer to Miners Ridge.

The Suiattle River corridor again became a focal point when reclassifying the Glacier Peak area. With its preliminary planning in 1950, the Forest Service proposed excluding the north side of the river from wilderness, believing it "very likely that these mines will develop into commercially producing properties," making a road for hauling ore out of the mountains "inevitable."[13] By the mid-1950s, a new company was operating at Miners Ridge, the Glacier Peak Mining Company, having tested in the area for a decade. The company's geologists remained "very enthusiastic" with the results of their annual summer work, anticipating plenty of copper and molybdenum, along with a little gold and silver mixed in, according to a Forest Service report.[14] Also seeing this future, the Mt. Baker National Forest supervisor wanted to exclude timber, facilitate developed recreation, and encourage mining and proposed in 1956 to split and reduce the Glacier Peak Limited Area, roughly at the Suiattle River. North of this would be a Snow King Wilderness Area; south of it would remain Glacier Peak.[15] The proposed Suiattle River Road had become a wedge, a tool to split the limited area.

In making these recommendations, foresters relied on myriad judgments about value. Those who prepared this report on the Mt. Baker National Forest certainly justified the economic value of timber.[16] By shrinking the limited area overall and excluding the valuable timbered valleys, forest supervisors estimated they would add almost 507 million board feet available for commercial harvesting. They figured this shift would increase the annual available cut almost 10 percent and add roughly $2 million in additional wages with four hundred more employees, a forecast that favored reducing protected acres. On the other hand, recreational land use lost money, from the agency's perspective. The report asserted that the Glacier Peak Limited Area could add $5.8 million to the economy through logging, while hikers and campers only contributed roughly $7.50 per visit. In the foresters' logic, three thousand people visiting the limited area annually cost the economy $1,934 per person—minus $7.50 each. Backpackers dragged the economy down. Even in recreation planning Forest Service managers could not shuck off their emphasis on the greatest good for the greatest number, titling a section of the report "Recreation Productivity."[17] Timber and recreation values were incommensurate.

Mining also influenced how the Forest Service redrew boundaries, and the agency used another set of values here. Perhaps because mining law prevented the agency from exercising much oversight and cash receipts did not flow back to it, the Forest Service offered no economic analysis of potential mines in the limited areas. In the mid-1950s Glacier Peak Mining remained optimistic that the increasingly scarce copper-molybdenum ore would pay off, and the company planned for an eleven-mile road to be built from the Suiattle River Road to Miners Ridge. The USFS saw a bright side to this development, with foresters writing that such a road would "open the flood gate to intense recreation pressures on the lower Suiattle River" and "be very heavily used."[18] With the agency's "greatest good" mind-set, this road would be a boon for the mining company and for increasing recreation in the heart of Glacier Peak country. Wilderness advocates, of course, wanted neither of these.

After the Mt. Baker National Forest offered its report, Region 6 issued its own, *Glacier Peak Land Management Study*, in 1957.[19] The Mt. Baker report had presented wilderness as a "departure" from the agency's history of multiple use.[20] By contrast, the regional office's study presented wilderness as consistent with Forest Service policies and practices, while simultaneously setting wilderness and its users apart as requiring special accommodation. Consider this characterization of a wilderness area: "It is a special type of recreation area catering to special classes of recreationists who possess the physical energy to hike or climb over rugged terrain, or who are financially able to hire pack and saddle animals to travel in the area. In the case of Glacier Peak, as well as in other wilderness area proposals, the Forest Service must consider the need for providing recreation opportunities for all classes of outdoor recreationists." In addition, recreation figures reflected the agency's prime value of maximizing returns and showed wilderness wanting. In the Mt. Baker and the Wenatchee National Forests, wilderness constituted 12 percent of the lands, but only 2,875 recreational visitors used those areas annually, half a percent of all recreationists. Although the report averred that wilderness values could not be quantified, it asserted its maximizing "objective . . . to promote greater use of all national forest recreation areas." So the agency hoped to punch the Suiattle River Road another six miles beyond its then-current terminus, up to Canyon Creek. This would help recreationists in two ways. First, the Forest Service might then develop campgrounds and other facilities along the road to serve the anticipated twenty thousand new visitors enjoying the forest every year and frequenting the small town of Darrington on its western perimeter. Second, it would get hikers that much closer to main features like Miners Ridge, Lyman Glacier, and Image Lake. The lake was always the area's marquee attraction. As the Region 6 report noted, "A visit to the proposed wilderness area could never be complete until the visitor has seen Glacier Peak mirrored in the calm waters of Image Lake."[21] Overall, the report emphasized wilderness as something special but perhaps extraneous; the agency needed to remember and plan for others who did not demand such catering, leaving an impression that wilderness users were being specially accommodated at some modest cost to the national forest system and local communities. As the agency continued its wilderness planning, it kept excluding the river corridors, allowing roads and timber sales to make deep gashes up the valleys, and this left the Forest Service vulnerable to criticism.

In February 1959, while deep snow covered all the North Cascades, regional forester J. Herbert Stone announced the agency's proposed Glacier Peak Wilderness Area, culminating nearly a decade of study and planning. Circulating the proposal widely, Stone informed interested parties that public hearings would follow in mid-October, just as snow would be returning to the mountains.[22] Like any story, the meaning of the proposal depended on how it was framed, and it presented a case study in how much can be said and how much obscured in a few short pages and a map. The agency sang the praises of the wilderness and its diverse "virgin forest types," glaciers, alpine meadows, high mountains, clear streams, abundant wildlife, and world-class trails. The proposal also emphasized the need for the Forest Service to balance commodity production and recreational aspirations. It hinted at sacrifice, an attempt to show flexibility and compromise. Fourteen percent of the wilderness contained merchantable timber. Four percent of that grew on unstable soil or was otherwise inaccessible to the market, but the remaining 10 percent contained 1.8 billion board feet of commercial timber, which according to Forest Service calculations amounted to thirteen million board feet it would forgo harvesting. Its final balancing act concerned recreation. Although wilderness recreation appealed to a growing number of mountain climbers and backpackers, the Forest Service's bread-and-butter recreationists favored developments such as roadside campgrounds. Foresters concluded aptly: "The proposed wilderness area boundary . . . is based on a careful analysis of all resources and uses of the general locality, and best estimates of the long-range requirements of all segments of the American public. Full consideration has been given to the protection and use of the area, as well as to coordinating wilderness values with existing and forthcoming developments. Establishment of the Glacier Peak Wilderness Area is in harmony with the multiple-use concept of national forest management."[23] The central themes the agency embodied—or at least aspired to embody—were there: balance, long-term planning, coordination, analysis, and harmony.

The controversial bit of the proposal, the Forest Service knew, would be excluding the Suiattle River, Chiwawa River, and Railroad Creek river corridors. As much as foresters paid lip service to undeveloped recreation, other priorities meant creating "deep indentations" up the three river corridors. According to the regional forester, these exclusions from the wilderness area would "greatly increase roadside recreation" and, as he put it in his statement at public hearings, "give easier and quicker access" to wilderness. Likely more important than faster wilderness access for those recreating in the forest would be how the road would "permit access to patented mineral properties," namely Miners Ridge. Smaller exclusions up the Agnes Creek and White River corridors also existed for "opportunities for roadside camping and picknicking," although timber harvest mattered in all these valleys.[24] These "indentations" created an oddly shaped map with fingers of unprotected land jabbing into the heart of the wilderness, leading conservationists to dub it the Starfish Proposal.

Although ostensibly for gaining public input, hearings held in Bellingham and Wenatchee, Washington, served a different crucial purpose for the Forest Service. They gave regional forester Stone an opportunity to explain the proposal and try to persuade the public of its wisdom. Stone rehearsed the merits as he saw them. He expected much higher revenue from recreation than in the past—another $50,000 annually in spending—by declaring a wilderness. His certainty of future mining gave him confidence that excluding areas from wilderness for roads would boost recreation, perhaps bringing in up to $250,000 annually to local gateway communities such as Darrington.[25] Forest Service officials consistently tied together extraction, wilderness recreation, and developed recreation, seeing them as inextricably bound and mutually supporting, a framing that conservationists opposed.

Over the decade while the agency redrew boundaries in the woods, conservationists developed their own framing. Concerned northwesterners gathered in organizations dedicated to enjoying and celebrating and protecting their favorite mountains. As the Forest Service was figuring out how to understand wilderness, scenery, and outdoor recreation, conservation organizations were preparing how to influence the outcome and respond to the agency's proposals. These conservationists formed a critical bulwark in the mountains to ensure that the Forest

Service reflected their views in its management plans. Twinned processes were underway—one by a public agency, one by the public *exercising* agency.

These conservationists had roots in Alpine clubs that had proliferated nationwide at the end of the nineteenth century, when climbing became popular. These clubs offered a way for men and women to test their strength, see high, wild country, and develop conservation ethics. This often eventually moved them into political campaigns to defend the places they came to love.[26] The first such club in the Pacific Northwest was the Portland-based Mazamas, started in 1894.[27] The Mountaineers launched a Seattle counterpart in 1906, after a Mazamas expedition on Mount Shuksan in the North Cascades. Charter members, roughly equal numbers of men and women, focused on outings and pioneering ascents in the Cascades. However, the Mountaineers included dedicated conservationists intent on preserving mountains unimpaired. The first president of the club wrote in his 1907–8 report that the Mountaineers planned "to render a public service in the battle to preserve our natural scenery from wanton destruction." By the late 1920s the Mountaineers was calling for a national park for Glacier Peak, but its political orientation weakened through decades of depression and war, reviving later. More than anyone, Polly Dyer made the Mountaineers surge on the postwar conservation scene, launching the club's conservation committee in 1954.[28]

Although many worthy places attracted the club's attention, the North Cascades and Glacier Peak in particular drew the attention of Dyer and other members of the Mountaineers, and they built a public campaign for wilderness there. Laura and Philip H. Zalesky of Everett, Washington, knew Glacier Peak well and showed it off through guided hikes and in print.[29] In a 1955 article in the *Mountaineer*, "Glacier Peak Area: Wilderness or Waste?," Philip Zalesky asserted that wilderness developed "soul and physical strength together," a perspective shared with Bob Marshall and others. Visiting Glacier Peak, he wrote, promoted this spiritual growth, with its "sublime . . . alpine gardens" and rare wildlife. On a recent trip sponsored by the Federation of Western Outdoor Clubs, Zalesky said, he hiked with Irston Barnes, a nature columnist for the *Washington Post* and president of the D.C.-area Audubon Society, who was "astounded" by the Cascades and whose presence demonstrated a growing national interest in this place. Sensing the importance of timing and audience, Zalesky laid out for the Mountaineers members the history of protection at Glacier Peak, emphasizing Marshall's expansive vision and the Forest Service's repeated reductions, citing the area's mineralization. As Zalesky put it, since mining laws "take precedence over wilderness areas . . . no part of a wilderness can remain sacred." "This being so, when the final decision is promulgated [for the Glacier Peak Wilderness Area], the boundary should include all the original 600,000 acres that remain in wilderness condition regardless of any previous conjectures made concerning ore bodies. High concentrate wilderness values should be allowed to stand on their own merit." If a mine was patented, then ingress and egress could not be prohibited, but there was no reason to preemptively exclude land, just as Ferdinand Silcox, the former chief of the Forest Service, had reasoned.[30]

While Zalesky's article represented the thinking of the Mountaineers and reports on political conditions regarding Glacier Peak wilderness, it did not emphasize something intangible but important about the trip. Joining the Zaleskys on this 1955 Cascades tour was Polly Dyer. After surveying wilderness on the eastern slopes of the Cascades, they waited for the ferry on Lake Chelan to take them back to the road home. While waiting in Stehekin, they met Jane McConnell, who insisted that they meet her husband Grant, the Berkeley-based political scientist whose summer residence there since the 1940s had inspired him to turn to conservation as both a scholarly subject and a personal commitment. The connections continued. Dyer's husband, John, had climbed Sierra Nevada peaks with David Brower of the Sierra Club, and McConnell knew Brower too. This knot of conservationists became crucial in publicizing the North Cascades and rallying for wilderness protection.[31]

Personal acquaintances such as these brought new people into the mountains and gave them firsthand experiences that changed perspectives; the movement developed experientially, as leaders of the Sierra Club had long known. Early in the club's history, outings to the Sierra Nevada were a primary recruitment tool. They were a brainchild of William Colby, who became secretary to club president John Muir in 1900 and served on the board for forty-seven years (and as president for two). The outings made the mountains better-known and popularized the Sierra Nevada and mountaineering as a communal recreational activity.[32] But the high trips became so successful that by the 1940s club leaders began questioning the ethics of large parties ascending to mountain passes and staying for a week or two. Too many in the mountains, with the pack animals to support a large commissary, damaged mountain meadows and interfered with solitude. Nevertheless, these trips remained popular.[33] By the mid-1950s, with North Cascades and Glacier Peak conservation campaigns pressing, the club sponsored multiple trips there.

A club trip to Glacier Peak in 1956, for instance, served multiple purposes. For the Sierra Club, trips into a region other than the Sierra Nevada helped spread its message and strengthened its ties with other conservationists. To counter emerging threats to wild places of the Northwest, the club had established a Northwest Chapter in 1954.[34] Surely Sierra Club trips to the North Cascades also helped make Californians—always the strongest part of the club's membership—aware of worthy landscapes outside their state. And such trips brought leading conservationists together. On the 1956 trip, for instance, Howard Zahniser, the executive secretary of the Wilderness Society, and his family joined David Brower, the executive director of the Sierra Club, and his family.[35]

The occasion provided time for Zahnie, as friends knew him, and Brower to savor their success in stopping a dam proposed for the Green River at Echo Park in Colorado's Dinosaur National Monument and to strategize about the wilderness bill that would soon become the prime focus of the organizations until it passed as the Wilderness Act in 1964. But family time in the mountains also offered welcome respite from hectic travel and meetings in boardrooms.[36] A color photo of a camp at White Pass that appeared in the *Sierra Club Bulletin* the following year shows a large family affair. A father and son walk in a meadow, with Sloan Peak towering in the background; a mother stands above a small child rolling in the grass; and clotheslines, tarps, and more than a dozen buckets appear—some of the gear needed to serve seventy-five campers.[37]

From these large outings publicized in the *Sierra Club Bulletin* to smaller backpacking trips led by the Zaleskys, enthusiastic visitors came to know the Glacier Peak region. Those with a literary or organizational bent then turned to publicizing their experiences and the threats they perceived, developing plans and narratives for protection. John F. Warth, a Northwest activist, added to this growing genre in *National Parks Magazine,* a publication sponsored by the National Parks Association, a private organization dedicated to keeping national park standards high. Sticking to conventions, Warth celebrated the various ways Glacier Peak and its environs offered an "unequaled," "unparalleled" wilderness experience. He touched lightly but critically on logging in the White Chuck River valley that opened views of the "ice-sheathed fire-peak" (Glacier Peak) that constituted the area's climax. But the real problem, the one Warth warned readers most strongly about, concerned Miners Ridge, the best place to view "this most elusive of volcanoes in its entirety." The ridge was endangered. Exploratory mining work already disrupted the once-tranquil area "with its helicopter port, tent camp, diamond drilling and unsightly mine tailings." But this intrusion was small compared with what Warth imagined in a "huge mining operation" with "all the sights and sounds of a modern mine—the shanties, the clanking mills, the groaning of truck engines, the dynamite."[38] The future lurked ominously.

Warth, who would go on to champion other Cascades wilderness areas, anticipated a mining operation splitting the proposed wilderness area—just as the local Forest Service supervisors once suggested—into separate northern and southern sections, but he understood that such a cleaving would ruin any protection that remained. All the trails met at Miners Ridge, he pointed out, and "it would be almost impossible to plan an extended hike that would not include this sober reminder of the ugliness of our industrial age." Here in *National Parks Magazine* Warth personalized Glacier Peak. Urging the public to stay apprised and participate in public hearings, Warth called for the Forest Service to preserve more than just mountaintops. Wilderness, he reminded readers, "is essentially a natural unit including the entire biota of an area. Remove any component, such as the lowland forests, and the wilderness is incomplete."[39] As the USFS began to reclassify Glacier Peak Limited Area, Warth offered perspectives that focused directly on major issues. Mines created ugliness. Wilderness encompassed the entire biota—it was a biological, not just a political or administrative category. This landscape merited national attention and protection. Any downsizing or severing during reclassification spelled the end of wilderness.

Northwest conservationists stepped up organizing efforts when the Forest Service simultaneously released its Glacier Peak plan and its decision to remove timber from the Three Sisters Primitive Area. The Mountaineers worked hard to stay abreast of USFS decisions, the new Sierra Club chapter stood ready to help, and writers like Warth and Zalesky publicized developments. But others, led by Grant McConnell, believed a new organization was merited, one with a single-minded focus on North Cascades wilderness. Modeled on Friends of Three Sisters and the Olympic Park Associates, the North Cascades Conservation Council formed in 1957, its name shortened almost immediately to N3C. Phil Zalesky presided, with Brower, Dyer, McConnell, University of Washington biochemist Patrick Goldsworthy, and others joining the board. The organization opposed a planned trans-Cascades highway, new dams, reduced wilderness areas, and any mine opening at Miners Ridge.[40]

Between 1957 and 1960, as Forest Service personnel devised and revised wilderness plans for Glacier Peak, Northwest conservationists activated the issue through their networks and publications. When, after years of study, the Forest Service finally proposed in 1959 to designate a Glacier Peak wilderness area and held hearings in Bellingham and Wenatchee, the backpacking public and others who cared about wilderness were primed to let the Forest Service know their preferences and criticisms. Nearly a thousand people spoke or wrote for these public debates, and the Department of Agriculture reported that the majority favored wilderness status.[41]

The hearings made a difference, at least to chief of the Forest Service, Richard E. McArdle. As he prepared to announce the Forest Service's decision on Glacier Peak Wilderness Area, he paid closest attention to the debates surrounding the river corridors. Most comments focused on the exclusions there, and that attention persuaded the Forest Service against opening all the valleys. McArdle said of the Suiattle that logging "would detract materially from the wilderness environment . . . and would seriously impair the view of Glacier Peak from such important vista points as Image Lake and Miners Ridge."[42] The testimony in Bellingham and Wenatchee had not moved regional forester Stone, who felt pressure to serve timber's booming economy and powerful political appetite. But the testimony moved McArdle or at least gave him visions of future headaches he wished to avoid. In the draft announcement he sent to Stone, he explained that a consensus indicated that the cuts were too many and too deep up those corridors. The Forest Service had decided to not exclude so much wilderness land up the Suiattle, Agnes, and Chiwawa drainages.[43] In other words, McArdle stopped the roads' extension. In the explanatory memo to Stone, McArdle also singled out the potential mine at Miners Ridge. With an eye toward long-term stability, Stone thought it foolish to declare something wild where there was a high possibility of a mine, but McArdle would not be cowed by a future mining complex. "If the mining properties in the Suiattle are ever developed commercially the Forest Service cannot stop access roads, townsites, and the degradation of the wilderness," the chief wrote. "Meanwhile, it seems unnecessary to anticipate such developments to the extent of refusing to dedicate the area as wilderness because of potential acts of people outside of Forest Service control and responsibility." Out of earshot of local interests, McArdle's position—in Washington, D.C., not Portland or Bellingham—ignored potentialities and probabilities and dealt only with conditions on the ground in 1960. And without an operating mine, there was no reason to accommodate one by creating an exclusion and thus allowing a route to it. Just as it had been during World War II, the Suiattle River corridor remained protected again, for the time being.[44]

So, in September 1960, the Forest Service established a wilderness area of 458,505 acres, a size that had expanded from a 350,000-acre limited area. Four years later, when Congress passed the Wilderness Act, Glacier Peak Wilderness Area became one of the first entries in the National Wilderness Preservation System. In simple terms, wilderness won, but it would become clear that the victory was partial.

With Glacier Peak Wilderness Area finally created, conservationists turned their strategizing to the larger region of the North Cascades and to strengthening the victory by solidifying the region's wilderness narrative. One victory was to be savored but could not mark the end of the effort. Another threat, another jeopardized wilderness, always stood ready and needing attention. A dynamic emerged organically in which local activists and national organizations cooperated. In 1964, in an article in *Ladies' Home Journal,* Supreme Court justice William O. Douglas, who had become a leading voice for conservation, called these arrangements "committees of correspondence." He took the label from the Revolutionary War, during which colonists corresponded with revolutionaries near and far to strategize and inform each other's work for independence. Douglas argued that conservationists likewise needed to coordinate efforts: "Local groups need national assistance; and that means joining hands in an overall effort to keep our land bright and shining."[45] Locals understood threats best but had fewer resources than, say, the Sierra Club.

More than anyone, David Brower had transformed the Sierra Club from a California hiking club to a national political organization. At the invitation of Grant McConnell, Brower turned his attention and knack for innovative publicity campaigns to the North Cascades and with a few others made a film in 1957, *Wilderness Alps of Stehekin,* to highlight the region. In the film, Brower narrated over scenes of hiking in high country, as beautiful as anywhere, emphasizing its accessibility by showing conservationists' children traipsing along alpine streams and trails.[46] The film earned awards and showed the North Cascades to perhaps hundreds of thousands of Americans. Polly Dyer remembered arranging more than one hundred showings herself.[47] In the 1950s conservationists used films to take people virtually into the backcountry, and their audiences included members of Congress. Brower and Howard Zahniser of the Wilderness Society, whose daughter was featured in *Wilderness Alps of Stehekin*, used films of Echo Park in their campaign stop the Bureau of Reclamation from damming the Green River in Dinosaur National Monument.[48] Shown in the halls of Congress and church basements alike, these films influenced conservation campaigns by increasing the number of people exposed to the landscapes—and threats to them.

Print media also represented a key part of wilderness conservation campaigns. Before moving to the Sierra Club full time, Brower worked as an editor at the University of California Press. Besides meeting his wife, Anne Hus, there, Brower learned valuable editorial and production skills that he took to the Sierra Club. Long a member of the club, Brower volunteered in 1946 for editorial duties at the *Sierra Club Bulletin* and promptly turned the publication toward more political aims. Later, after the Sierra Club sponsored a successful Ansel Adams photo exhibit, Brower thought the experience might "be bottled" in a book. Thus was born the Exhibit Format book series, with Adams's *This Is the American Earth* the first title.[49]

The Exhibit Format books were unusual. For one thing, they were large-format, 10.5 inches by 13.5 inches. They were also costly. The first one was priced at $15—$120 today—and soon they were $25, roughly the equivalent of $200. Despite this exorbitant cost, they sold well, suggesting a ready audience among well-to-do Americans. Their first title with color photos (and the higher price), *"In Wildness Is the Preservation of the World,"* ultimately sold around a million copies. The Sierra Club books were also unusual in that they merged genres. They were part art exhibit, part nature writing, in an award-winning blend. The Exhibit Format books were beautiful, their writing was eloquent, and they conveyed the Sierra Club's conservation message clearly, with all its politics. Brower intuited that words and images combined could influence the public; the Exhibit Format series bore out his instinct.[50]

Thus it was not surprising that the North Cascades eventually drew attention for what became the eleventh book in the series. In 1965 Brower turned to Harvey Manning to write *The Wild Cascades: Forgotten Parkland.* An activist whose trail guides still find their way into backpacks, Manning enlivened Northwest conservation circles for half a century with his piercing prose. He tried to tame that prose for *The Wild Cascades,* once telling a correspondent that a revised draft was "somewhat milder, or at least less wildly speculative about what goes on in Forest Service minds."[51] That the book even made it to production was a testament to Manning. Brower asked him in March to write the manuscript for publication in June. Only a fool would agree to such a rushed schedule, but Manning possessed sufficient passion and fortitude to charge ahead. He also had been living and breathing this fight with N3C, so the story and the campaign's contours were clear to him and could be written quickly. Manning collaborated and incorporated others' ideas, responding to one who suggested revisions on an early draft, "Certain phrases, sentences, quotations, etc. I've taken over intact, so that what we now have is the Goldsworthy-McCloskey version, as amended by Marshall, and as passed through the Manning typewriter."[52] (Patrick Goldsworthy led N3C; Michael McCloskey served as conservation director for the Sierra Club; and George Marshall was then president of the Sierra Club.)

When published, the book included words from Brower, a foreword from Justice Douglas, text by Manning, photographs from the likes of Philip Hyde and Ansel Adams most prominently, and excerpts from poems by Theodore Roethke.[53] Such a roster for such a cause in such a publication guaranteed widespread attention at the very moment when conservationists transitioned from the triumph of passing the Wilderness Act to securing that victory.

Douglas set the tone in *The Wild Cascades* with his opening words. Like many, Douglas hearkened to the frontier, seeing it as a fount of national identity and pride. This trope went back to the popular histories by Theodore Roosevelt and more academic treatment by Frederick Jackson Turner before 1900. These men argued that the frontier made the United States exceptional. Such a historical interpretation amounted to myth-making and excluded many people, notably Native people who made a home out of what many called wilderness. But the idea that America's continental expanse and uncommercialized spaces made the nation distinct resonated well into the mid-twentieth century. Indeed, it helped form a major interpretation of American wilderness, an interpretation in historian Roderick Nash's classic *Wilderness and the American Mind,* first published in 1967—a book Douglas reviewed favorably.[54] In his foreword to *The Wild Cascades,* Douglas warned, "If we do not preserve the remaining samples of primitive America, we will sacrifice traditional American values, the values of frontier America. . . . As long as that continues we will retain a historic connection with the past of our nation—and our race." Given his long record in favor of civil rights and the broader context of his voluminous writings, it is clear that Douglas meant the human race, although the phrase sits as an uncomfortable reminder that wilderness conservation remained a predominantly white movement. Douglas foreshadowed the book's call for action as his foreword closed. "We need a number of protected wildernesses along the Cascade range. . . . But we also need—and most of all—a North Cascades National Park. And that's the special message of this book."[55]

That was indeed the message. Manning made it personal and political, and he nudged into the polemical. Manning drew from his earliest memories of camping in the North Cascades and fishing with his father on the North Fork of the Stillaguamish, just a drainage over from the Sauk and Suiattle system. He had repaid his debt to his father by taking his own children to the mountains when they still wore diapers.[56] In *The Wild Cascades* Manning took readers through short chapters extolling the virtues of "rain sleep," timbered valleys, and transitions to summits and passes. Masterful full-paged black-and-white photos showed cedars with branches so sweeping no trunk could be seen and alpine lakes so nestled that they seemed safe from any possible invasion. Manning knew the whole range but seemed especially fond of the area centered on Glacier Peak. It offered much and was easily accessible from Seattle, serving as the gateway to the entire North Cascades. And his prose was evocative and poetic. Consider a typical passage:

> On another bright and windy day one may climb Miner's Ridge through red heather, white heather, and yellow heather, all in fresh bloom, all mixed together, and the slope so steep that the red bells, the white bells, the yellow bells are only inches away from eyes and nose, and at length one seems not to be climbing upward on feet but swimming upward with hands and knees and elbows, affecting a butterfly stroke through a multi-colored froth of silent bells.[57]

Such descriptions transported readers to an almost otherworldly place where hikers flitted like butterflies through mountain flowers as colorful as rainbows. With photographs to prove such words were no flight of fancy, *The Wild Cascades* rendered the mountains as a fantasy landscape that was true.

If the mountainscapes were not fantasy, they were fantastic. Manning wanted readers to understand the nature of the North Cascades' wildness. Belying charges of elitism, the likes of which appeared in earlier Forest Service reports, its diversity was not just for the "heavily muscled, highly skilled, and fearless. . . . There are many varieties of wilderness experience, and no absolute calculus by which one can be ranked above another, except in the degree of genuine wildness, as measured by freedom from the sights and sounds of machinery and sheep and other gross intrusions of civilization, and by the unrestricted, uninterrupted continuity of nature and its processes." For wilderness advocates, these qualities were paramount: a continuation of natural processes, including the wildlife he highlighted, from cougars to spiders. Manning filled the book with seasons and sights. He continued the theme of wilderness essences to conclude a chapter: "There are many varieties of wilderness experience, and North Cascades travelers may argue among themselves which is the best. However, all agree that this is a wilderness which must be preserved undiminished and fully natural, for their sakes and for those of their children and great-grandchildren, and for all others who seek occasional refuge from the pressures of a steadily more raucous civilization."[58] With such characterizations, full of long-range thinking and commitment, Manning ended the celebratory part of the book, which consisted of nine two-page chapters.

Two chapters remained—"Nibbling" and "Toward a Park"—in which Manning politicized the place, launching a campaign that would punctuate the North Cascades' history. Manning walked the fine line, often present in wilderness issues, between the power and the fragility of nature. "For all the power of its rivers and storms, the immensity of its peaks and the ancience of its forests," he wrote, "wilderness is a delicate and dynamic balance between many forces, including a natural succession of forms of life; the balance is all too easily upset and the wilderness destroyed." With Glacier Peak protected as official wilderness, the North Cascades more broadly remained vulnerable, its balance poised for upset. Manning reported that multiple-use management—"the doctrine that Forest Service land managers can produce a cornucopia of economic and recreational riches without disrupting the natural balance or reducing the wild beauty"—was knocking at the Cascades' door. He knew that both wilderness and multiple use inevitably would remain among the peaks, but he strove to help tip the balance toward wild scenery and not roads, vistas without foregrounded stumps, and quiet hikes undisturbed by "the sounds of bulldozer and chainsaw."[59]

While Manning railed longest and most fiercely against the timber industry (and the Forest Service's abetting of it), he introduced readers to the problems of mining, including subtle problems that plagued the North Cascades and that national park advocacy meant to solve. On the east side, in the Railroad Creek drainage just over the crest from Miners Ridge, sat an abandoned town: Holden. The valley had been poisoned after a mere fifteen years of active mining and provided evidence of how quickly destruction comes with certain activities. Manning considered the entire North Cascade range vulnerable to "the whim of the corporation miners" then "prospecting the range on foot and in helicopters" at Miners Ridge and beyond. The Forest Service could not override the Mining Act of 1872, a "decrepit law" in Manning's words, even in places protected by the new Wilderness Act. Activists from the Sierra Club would have known that, but a casual reader who picked up the Exhibit Format book for its photos might have been surprised. Manning presented a solution: "only in a properly dedicated national park can miners be made subservient to the larger national interest." Manning and his allies had concluded that creating a national park was the only way "to halt the continuing nibbles and gulps of multiple use."[60] The Forest Service, unable to conceive of wilderness preservation in grand enough terms and shackled by old and new legislation, proved to Manning and associates that the agency's limitations were too great.

Manning offered a history of the National Park Service and the Forest Service in these mountains, a history that emphasized the threat of the former to the latter's priorities. The Forest Service reacted to losses of landholdings and the popularity of the national parks with programs meant to demonstrate its wilderness bona fides. When it had designated Glacier Peak in a so-called recreation unit in 1926, some 233,600 acres, the local forest administrators were bulwarking against the Park Service making inroads. Competing proposals during the 1930s gradually increased protected national forest land, but well into the 1950s the Forest Service remained a reluctant and ambivalent conservator of the North Cascades wilderness. And this reticence eventually convinced regional conservationists to throw in with the Park Service.[61]

The North Cascades Conservation Council led the way. Its proposal in 1963 for a park earned endorsements from the Sierra Club, the Mountaineers, the Mazamas, the Cascadians, the National Parks Association, the Wilderness Society, the Federation of Western Outdoor Clubs, and more. The proposal limited roads and outlined a wilderness park, except for the North Cross-State Highway (now Highway 20), then too far advanced to be stopped. The park would restrict commercial activities within its boundaries. To accommodate longstanding hunting traditions, N3C proposed an adjacent Chelan National Mountain Recreation Area, where hunting could continue in a setting every bit as scenic as the national park. In *Wild Cascades* Manning countered vocal objections at the time, pointing out that the annual take of deer within the park's proposed boundaries amounted to 1 percent of the state's typical harvest. He acknowledged the loss of logging and related jobs that would result, some 314 jobs, but anticipated nearly 2,100 new tourist industry positions by 1980. Finally, he reminded readers that the merits of the proposal ought not to be based on simple economic calculations but also on the consistency of the North Cascades with national park standards—standards the area more than met.[62]

Tucked inside the back cover of *Wild Cascades*, a detailed map showed the boundaries proposed for the park and recreation area. It was expansive—and not just because the map measured three feet by two feet when unfolded. The boundaries of the national park stretched from the Canadian border southward to encompass the Glacier Peak Wilderness Area. The dream borders projected, Manning and his substantial stable of allies redoubled and refocused their efforts. A decade before, the North Cascades had enjoyed no permanent protection of its wilderness, only scattered administrative guardianship. Manning's book, an underappreciated part of the campaign, solidified the region as a national conservation priority and helped it gain attention beyond N3C and similar organizations.[63] Evidence that the book worked as intended came from eastern Washington's *Wenatchee Daily World*, a paper that had long championed fully developing natural resources. The paper characterized Manning's work as "irresistible . . . propaganda ammunition" and worried that Congress might shut the North Cascades off from any future development beyond tourism, which most extractive-focused mind-sets did not view as a viable economic future.[64] Even if Manning had not convinced the *Daily World* of the merits of a nearby national park, he had persuaded its editors of the effectiveness of his prose.

While the *Daily World* feared that *The Wild Cascades* would work as propaganda, the president of the Sierra Club praised it. George Marshall, whose brother Bob had moved the North Cascades toward preservation in the 1930s with his passion for wilderness conservation, wrote to Manning to express his gratitude. He told Manning that he had recently flown over Southern California and allowed the urban sprawl's "ugliness in many crippling forms" to depress him. But, arriving home, Marshall found Manning's book, which succored his spirits. "Wilderness does not solve all problems by a long shot," Marshall philosophized, "and for some people it solves nothing at all, and yet without it and other forms of beauty and wildness, life would be come [*sic*] drab and closed-in and probably would disintegrate, not only for many of us who are aware of its need to us, but to humanity as a whole." Marshall drew out the essential role that wilderness played in modern times. He acknowledged that a protected North Cascades wilderness might not seem to solve the problems of "prejudice, areas of deep poverty, ghettos" that he flew over, but he argued that such wild places do serve "humanity as a whole." Marshall praised how Manning captured "the theme of a love of these mountains and the enjoyments of trips in them to an effective, passionate condemnation of those who would destroy this wilderness and of the means they will use if not stopped."[65] With this affirmation of Manning's book, Marshall captured much of the entire wilderness project. Wilderness was essential for humanity; it rested on love; it required action when its destruction loomed.

Protecting the North Cascades had occupied Northwest conservationists for a long time, but this moment saw them pivot. Grant McConnell announced in the mid-1950s that the North Cascades had been accidentally saved and urged a broader audience to pay attention and work for protection by deliberate action rather than happenstance. In 1964 Glacier Peak achieved the highest possible status: legislative wilderness through the new Wilderness Act. Still, conservationists saw their work there and the broader North Cascades as incomplete. They needed to tighten protections. The local and national campaigns for a national park became highly visible, but within the wilderness were minerals attracting attention.

No sooner had Congress declared Glacier Peak one of the inaugural wilderness areas than Kennecott Copper Corporation ramped up its exploration of Miners Ridge, having acquired the claims in 1954 and picked up earlier companies' exploration activities. The company wanted copper. The company wanted profits. But more than that, the company wanted to ensure that the Wilderness Act would not be used to keep it from exploring and developing mineral deposits there or in other wilderness areas. The North Cascades, then, stood as a proxy for a larger concern about resource extraction in the West. Glacier Peak was a test, a challenge to extraction's primacy and the strength of wilderness laws and advocates' resolve. The players were in place by the mid-1960s. The Forest Service and the National Park Service would compete. Kennecott would poke and prod. Conservationists would innovate in their commitment to confront institutional and political obstacles. The campaign to protect the Wild Cascades was heating up.

PART TWO
Challenges

THREE | The Dance

THEY DANCED WARILY. IN the mid-1960s, Kennecott Copper Corporation and the US Forest Service approached each other, forced partners, neither sure it could trust the other. Forest Service managers at several administrative levels stumbled forward in their attempts to take the lead, to control the drama of Miners Ridge, before finding assured footing. However, once a forester found his confidence, even that caused problems since not all personnel danced to the same tune. Harold C. "Chris" Chriswell, the Mt. Baker National Forest supervisor, brought with him a good deal of experience. During his career he bounced around Region 6 in Oregon and Washington, in forests on both the east side and the west side of the Cascades, where he learned how to manage grazing, timber, and recreation. In the 1950s he showed a willingness in certain circumstances to lower commercial harvest goals, controversial in those days when "getting out the cut" drove the agency's agenda and western Washington's economy. But if occasionally Chriswell might alter and slow timber production, he typically supported Forest Service practices and punched roads up river valleys to support harvests.[1] This ambivalence may well have influenced Chriswell's understanding of the Wilderness Act, which he thought chock-full of ambiguity. When Kennecott announced in December 1966 its plans to start an open-pit mine on Miners Ridge, the forest supervisor acknowledged that the courts might ultimately decide rules, but in the meantime he assured the public of the agency's power to "control things," saying, "we have told Kennecott it would have to bring in all possible alternatives if, as and when it makes a formal application. . . . We will exercise all control possible within the law to protect wilderness values."[2] Chriswell promised nothing specific, promoting wilderness values and asserting that he would control Kennecott as much as the law and courts would allow. Such public statements from Chriswell, as mushy as they were, might have encouraged conservationists. But they were neither the first nor the final words on the topic.

Like long-feuding families, Kennecott and the Forest Service danced around the politics as if at a community square dance, working to set aside their differences long enough to execute a grand right and left.

The Forest Service knew Kennecott was coming. In the months preceding the corporation's public announcement, the agency responded at least three times to activity or concerns related to Kennecott. Internal Forest Service correspondence reveals how foresters and agency officials wrestled with the Wilderness Act's implications as the company prepared to test the mining exception. One concern entailed Kennecott's use of helicopters; one focused on the compatibility of mining and wilderness protection; and one was the agency's plan to counter what it anticipated the company proposing at Miners Ridge.

Helicopters flying mining executives and engineers in and out of a wilderness violated most commonsense notions of wilderness. Special provisions in the Wilderness Act explicitly allowed prospecting as long as it was "carried on in a manner compatible with the preservation of the wilderness environment."[3] As early as January 1966, the corporation pushed the Forest Service to allow helicopter use. Without it, the company argued, prospectors could not be expected to survey wilderness areas for mineralized land before the act's special provisions ran out on December 31, 1983. Chief forester Edward Cliff, who had taken over for McArdle in 1962, responded that the Forest Service had decided that "if a needed activity could be accomplished by nonmotorized means, it would be," letting Kennecott know that the Forest Service held itself to that standard and made no exceptions for other government agencies either. Some also wondered how to determine whether using helicopters for prospecting "was actually bona-fide" or simply convenient and cheap.[4]

Forest Service personnel anticipated letters like the one a Mrs. William Devin wrote the following summer to Secretary of the Interior Stewart Udall complaining about helicopters landing a "short way beyond our camp" near Image Lake. Devin and her husband believed that this put private interest above public welfare and might be illegal.[5] Since Miners Ridge was in the heart of the well-used recreation areas of Glacier Peak, Image Lake, and the Pacific Crest Trail, the Forest Service could be dogged by many such public expressions of concern.

Despite stated standards and public complaints, helicopter landings continued and vexed the agency. The Forest Service informed the miners that they could land only at the main camp; instead helicopters landed at several locations. Local foresters sought assistance from those higher up and urged them not only to limit landing sites but also to specify appropriate travel routes. "Buzzing the campers at Image Lake is as much a trespass as landing at the area," one wrote.[6] In response, attorneys and high-level Forest Service administrators in the Portland and Washington, D.C., offices studied the ways the Wilderness

Act affected companies' special use permits and helicopter landing privileges. As might be expected with lawyers, bureaucrats, and myriad regulations, no clear outcomes emerged from multiple memorandums layered in closely defined rules and replete with caveats. In the end, the agency recommended that local foresters work closely with the mining company, hoping that might ensure all that regulations were followed. The correspondence showed that the foresters at Mt. Baker National Forest—the ones with the closest responsibility for managing Glacier Peak—thought that helicopter landings violated wilderness values and wanted to reduce and confine them. Internal memos recorded their strong desire to limit, even hamper, Kennecott's emerging plans. The uncertainty in upper administrative levels demanded a cautious and even accommodating approach as all sides stumbled forward testing the Wilderness Act.[7]

At the same time the helicopter question occupied foresters from Darrington to D.C., Northwest conservationists pressured administrators and policymakers to explicitly define the compatibility of wilderness and mining. NC3 director Goldsworthy anticipated Kennecott's formal announcement, seeing all the activity at Miners Ridge as making it inevitable, and he wrote to regional forester Stone, asking pointed questions and pressing for regulations. Goldsworthy sent copies of his letter to Chriswell, the Mt. Baker National Forest supervisor; Secretary of Agriculture Orville Freeman; Henry M. "Scoop" Jackson, Washington's powerful senator and chair of the Senate Committee on Interior and Insular Affairs; and Edward C. Crafts, director of the Department of the Interior's Bureau of Outdoor Recreation.[8] Each of these men investigated and sought clarifications from various parties from Portland to the nation's capital.

Spurred by information "from a usually reliable source," of "the imminence of a disaster in the North Cascades," Goldsworthy sought specific information and assurances from Stone. Goldsworthy's source indicated that power lines, a road, a mill, a town, and a tailings dump were all part of Kennecott's plan. The conservationist wanted to know from Stone whether Kennecott had requested permission for such developments. If so, or if the agency expected it, Goldsworthy wondered what "standards, limitations, restrictions, etc." the USFS might impose. Goldsworthy was particularly eager to know where power lines, roads, or mill sites might be, implying that these industrial artifacts ought to be located outside the forest. Because he worried about water pollution like that from the nearby Holden Mine, he also urged Stone to prohibit siltation and enforce this prohibition strictly. As if the point needed amplifying, Goldsworthy challenged the agency: "The Forest Service may now really be put to the test to see if it can really protect the Glacier Peak Wilderness Area by imposing stiff regulations on mining and rigidly enforcing them."[9]

Due to his growing reputation, Goldsworthy received answers. The responses varied, depending partly on the constituencies each correspondent served and because the issue had barely begun percolating in these wider circles. No one had devised straightforward answers yet. Crafts and Jackson, for instance, replied with general support for Goldsworthy's efforts, but they acknowledged the ways that the Wilderness Act limited options because of its specific support for existing mining claims and continued prospecting. Stone himself did not reply directly, tasking his regional office assistant, A. E. Spaulding, with it. Spaulding offered Goldsworthy slight relief, telling him that Kennecott had not requested any permissions yet. Since the company had shared no plans, Spaulding believed it "impossible" to answer the other questions and vaguely reassured Goldsworthy that the Forest Service would decide "within the framework of the applicable laws and regulations." If Goldsworthy had thrown down the gauntlet, the regional office stared at it almost mutely.[10]

The Washington office of the Forest Service had more to say, although probably little of it assuaged Goldsworthy and his compatriots. Secretary Freeman, responsible for the agency, assigned the task of replying to associate chief Arthur W. Greeley, son of William B. Greeley (Forest Service chief throughout most of the 1920s). Since Kennecott had not yet applied for permits, Greeley provided no specific information about the company's plans or the Forest Service's requirements at Miners Ridge. However, Greeley detailed the ways the Wilderness Act precluded the sorts of regulations Goldsworthy pressed for. The conservationist's challenge to the Forest Service irritated Greeley, who lectured Goldsworthy. The Wilderness Act permitted mining; there was no escaping that. "You know what the regulations are," Greeley said, adding, "The mining industry is not very happy about them." Goldsworthy's hope—to regulate mining away—was unreasonable. Greeley told him: "It is important that our position be clearly stated. The law provides for mining. We cannot nullify the law by writing general regulations that would make mining not possible. We have just as much obligation and desire as ever to maintain wilderness values. But we also have the obligation to administer the mining provisions of the Wilderness Act in a manner consistent with Congress' intent—which is that mining be permitted."[11] Greeley picked up the gauntlet and threw it back.

With only this evidence, Greeley might appear a pro-mining zealot, but ten days after he replied to Goldsworthy he wrote to Lloyd Meeds, the U.S. representative for the Washington congressional district that included Miners Ridge. Greeley shared much the same information and legal interpretation as he had with Goldsworthy, but he offered Meeds a slightly different perspective: "We concur with your thought that mining operations are not consistent with the maintenance of wilderness values. But the law provides for mining, and we have the obligation to administer the mining provisions as well as the other provisions of the act."[12] Here the associate chief of the Forest Service said more clearly than ever: mining and wilderness do not mix well, but we cannot help it. This sense of its incompatibility and the apparent frustration of having administrative hands tied legislatively continued to gnaw at the Forest Service as Kennecott prepared to make its move.

At last the agency prepared for Kennecott's formal proposal, an inevitability at this point. For months, if not years, the Forest Service, conservationists, and Kennecott employees all waited for the moment when Kennecott would stop testing Miners Ridge for ore and start mining it. By late September 1966, Mt. Baker National Forest supervisor Chriswell prepared for an upcoming meeting with Kennecott and contemplated ways to reduce Kennecott's impact in the wilderness. He intended to require the company to use block cave mining—a less intrusive but more expensive method—rather than open-pit mining and to segregate the townsite, mill, and certain processing activities outside the wilderness and even outside of the national forest if possible. In short, Chriswell searched for ways to minimize the mine's impact, suggesting that Kennecott be required to bury power lines and move ore through a pipeline. His was a strong attempt, but his legal advisors in the Portland office told him that he likely lacked authority to carry out this plan. Wilderness values, an attorney in the regional office reminded him, "must here be weighed versus costs." Such advice had little specificity but was fully consistent with the Forest Service's longtime efforts to balance everything. The attorney's final words, though, revealed the weightiness of the upcoming meeting with Kennecott representatives: "You may wish to submit this matter to the Chief since precedent-making decisions seem to be involved." The frank acknowledgment of this being new territory for the agency and the suggestion that the chief forester might need to weigh in demonstrated that Miners Ridge sat at the center of national management questions.[13]

Finally, on November 3, 1966, Kennecott and the Forest Service sat down together in Bellingham. Two men from Kennecott, along with two people from the engineering firm they employed, met with six Forest Service employees (three from the Mt. Baker office, two from the local ranger district in Darrington, and one from the regional office in Portland) and detailed their plans. Kennecott's vice president for new mines, M. J. O'Shaughnessy, said that the "stable price of copper and the excessive demands for copper" finally warranted the company putting a mine into production, pending the final outcome of a few more tests. Although the mine complex they envisioned was much smaller than the company's massive open-pit mining enterprise, the Bingham Pit outside Salt Lake City, Utah, it arguably was more intrusive. Near a ridgetop with a few intersecting hiking trails and a fire lookout amid alpine lakes, meadows, and glacier views, Kennecott proposed to dig a hole a thousand feet deep and sixteen hundred feet wide. Its waste would be held back by a dam 250 feet tall, nearly the height of a twenty-five-story building, plopped down in a mountain wilderness. These first specific details from the company seemed incongruous for wilderness.[14]

From the beginning, the company recognized the particular challenge the Wilderness Act imposed. More than half of the Forest Service's summary of the meeting focused on the mine's implications for wilderness. Kennecott was "very mindful of their public image" and wanted no damage to it, so its officials assured the USFS that they would "minimize the cultural effects" of the mine and abide by all the Wilderness Act's provisions. Much like the Forest Service, Kennecott recognized the precedence this proposal furnished. The company executives felt they were "representing the mining industry in this first venture, in particular, and [wanting] to demonstrate that mineral exploration and development [was] compatible with the Wilderness Act as it has been created by Congress." All parties saw it this way and knew much was at stake. The Cascade Crest Trail, which soon would be designated the Pacific Crest Trail under the 1968 National Trails System Act, would be "breached" by the pit, Kennecott officials acknowledged. The popular hiking trail would be rerouted above the mine, and the corporation's officials thought it would still be "fairly well insulated against the sound and view of the mining operation." Such a belief—that a small trail reroute could mask the sight and block the sound of five thousand tons of ore being mined and milled every day—amounted to magical thinking. On-the-ground practicalities intersected with legal requirements, as Kennecott knew full well. O'Shaughnessy promised to obliterate "cultural inroads" after mining was completed, in the "full spirit of the Act—that is, in a reasonable manner," a promise that left ample wiggle room. The meeting concluded without a formal request or decision, but the path seemed set: forward to the pit.[15]

The layers of this story deepened. Kennecott Copper had finally come out of the shadows to meet with the Forest Service, although it had yet to apply for required permits. The conservationists mobilized their allies and queried administrators and legislators now that something more than rumors and vague potentialities hung in the air. At the center of the story sat the Forest Service, but the agency spoke with multiple voices and not just because it spoke to multiple audiences. Different levels of the agency—district, forest, region, and nation—saw different opportunities and constraints and felt different commitments to the Wilderness Act. As Kennecott began solidifying its plans, it met an agency unsure whether to resist or acquiesce, and if it wanted to resist, did it have the authority to do so? Untying this knot requires knowing how private land and enterprise tied up the public lands in the first place.

When the Forest Service was created and placed in the Department of Agriculture in 1905, Gifford Pinchot became chief forester. Pinchot wrote his own marching orders and signed them with the name of his boss, Secretary of Agriculture James Wilson. The so-called Wilson Letter laid out the agency's direction, a mandate that would guide it for seven decades. Besides paeans to maximizing use, avoiding monopoly, and conserving resources—all Progressive causes in general and conservation causes in particular—the Wilson Letter gave primacy to the local: "You will see to it that the water, wood, and forage of the reserves are conserved and wisely used for the benefit of the home builder first of all, upon whom depends the best permanent use of lands and resources alike." And further, "In the management of each reserve local questions will be decided upon local grounds."[16] These policies prioritized community use of national forests, places where local ranchers ran their cattle and sheep, local mill owners bought timber, and local families camped. Grazing fees, timber sales, and recreational leases (and in later years use fees) helped fund the agency's management activities, although all were below market value and never matched expenditures. By providing access to public resources at subsidized rates, among other things, the Forest Service quickly reversed the local opposition that had initially appeared when federal bureaus asserted control over the public domain. Doing so evolved from nineteenth-century practices and policies of economic liberalism, a way for governments to facilitate resource development with minimal interference.[17]

In the nineteenth century, the federal government contrived myriad ways to give away land from the public domain. When that priority shifted toward maintaining some public land and conserving it, national agencies like the Bureau of Reclamation and US Grazing Service administered much of the West's water and land policy. Despite having a Washington, D.C.-based agency nominally in charge, interest groups pushed to decentralize the administration and instead rest power in local offices. This largely successful strategy minimized the strength of national bureaucrats and directives. The political scientist Grant McConnell cared deeply about how interest groups functioned in the American political system. He also cared deeply about the North Cascades, with his summer home in Stehekin perhaps thirty trail and fifteen air miles from Miners Ridge. In his classic study *Private Power and American Democracy,* published the same year Kennecott announced its open-pit mine, McConnell argued that local officials served local elites with near autonomy. What McConnell phrased as the "cult of decentralization"—another observer, more admiring than critical, called it an "ideology of decentralization"—described the ambition to reduce power, but in fact for private interests "power was enhanced by decentralization," because decentralization weakened the regulatory power of government administrators when public and private bodies disagreed. This important paradox animated much of resource agencies' histories: in being chiefly responsive to local demands, they became unable to assert necessary administrative functions against stronger corporations or other interest groups. This perspective certainly seemed to confirm McConnell's experience in the 1950s, when timber companies were eager to cut into sacred wilderness valleys like the Agnes Creek drainage, and history seemed ready to repeat itself with Kennecott in the 1960s.[18]

The Forest Service exemplified the general dynamic of resource agencies and politics, exceptional only in its longevity and abilities but typical in its orientation. Policy scholar Christopher McGrory Klyza said of this early period, "The Forest Service was an island of capacity and autonomy in a generally weak state."[19] That weak state assured decentralization, even as foresters on the island soon enjoyed respect from locals and, importantly, autonomy from the national office. As McConnell had earlier concluded, decentralization meant that power accrued to local elites; in this case local forest rangers worked at the behest of the more powerful local industries. After World War II, the national forests started serving corporate interests to a greater degree as demand for public timber grew.[20] Between 1950 and 1966, for instance, the Forest Service allowed more timber to be cut than had been cut in the preceding forty-five years.[21] Well before this massive new harvesting, the precedent had been set; private enterprise—the local ranch or local mill—long profited from using public resources. But ranchers and mill owners paid something through their grazing fees and timber sales.[22]

The policies that governed mining on public lands, like Mt. Baker National Forest, flowed from that same decentralized mid-nineteenth-century stream but gushed into the twentieth century with even fewer obstructions. Mining cost miners relatively little. The General Mining Law of 1872 culminated congressional interest (and continues to a large extent to serve as mining's legislative lodestar, despite the passage of more than fourteen decades). It opened all of the public domain to mining—"all valuable mineral deposits in lands belonging to the United States, both surveyed and unsurveyed, are hereby declared to be *free and open* to exploration and purchase"—and allowed anyone to claim public land for prospecting; no payment for the minerals once found was required. At the time of claiming it, miners needed to testify that they had invested $500 in labor and improvements, after which they needed to invest $100 per year in labor or improvements to maintain the claim. The land itself could be patented—that is, formally deeded from the federal government—for merely five dollars per acre. Miners could also work unpatented claims and extract mineral wealth but never acquire actual title. By any market measure, this system was a bargain.[23]

The conservation laws that passed from the 1890s through 1910s set aside millions of acres in national forests. But the 1872 mining legislation had precedence, so the Forest Service could not restrict prospecting or mining—national forest land remained "free and open" for mining. For timber the agency opened up tracts for sale, solicited bids, and imposed some guidelines; for mining the Forest Service did nothing. And nothing was required of prospectors, not even informing the Forest Service of their claims. By 1951 the Forest Service reported nearly a million acres of patented mining claims on national forest land, although fewer than 15 percent ever operated commercially. All told, patented and unpatented, more than two million acres of mining claims were scattered throughout the national forests, with less than 3 percent active in 1950, although these figures could only be estimates since no reporting was required. This array of practices and contexts gave Kennecott its claims: 300 patented acres and 2,650 unpatented acres. Pending studies and further exploration, Kennecott might have converted some of the unpatented land to patented acreage, where the

Forest Service possessed little regulatory power. Kennecott's private oasis amid the public estate was fully consistent with the prevailing belief carried out of the nineteenth century into the twentieth that public land and its resources ought to unleash the creative and economic energies of private enterprise.[24]

McConnell's personal and professional histories overlapped. What he called the "Progressive tradition of American politics" saw private groups' power as "excessive and dangerous." To counter this, Progressives worked to destroy such interest groups or reduce their power, often through government regulation. On the other hand, private groups played a vital role, and many believed that small organized groups were virtuous, essential curbs to government power. For McConnell, the party system, the presidency, and the national government functioned to tamp down narrow agendas and the privileges afforded to special interests, such as mining corporations. *Private Power and American Democracy* appeared just as Kennecott announced its interest in extracting copper from public land, but McConnell understood that only vital political engagement by citizens and a watchful Forest Service could effectively confront the corporation.[25]

Kennecott finally announced its plans in December 1966.[26] Conservationists had been prepped for it, but the Forest Service remained the most immediate obstacle. Conservationists could not anticipate the agency's response, and its history of reducing wilderness protection during reclassification set an alarming precedent. The Forest Service possessed few tools to challenge mining on its lands, and after the Kennecott announcement the agency continued to be ambiguous in its response, ultimately only succeeding in mixing signals. Mt. Baker supervisor Chriswell embodied this ambivalence: he assured the public that the agency would "control" the mining; he expected that the courts would resolve it; he hoped, vaguely, that the Forest Service could protect wilderness values. Ken Blair, the Wenatchee National Forest supervisor, told a man from eastern Washington that he did not approve of Kennecott's plans and encouraged him to spread the word to "your conservation friends" and to write Senator Jackson.[27] Meanwhile, the regional forester in Portland also spread confusion. Herb Stone had ascended to the regional forester position in 1951, before which he had overseen the southern region for the five years after World War II. Accordingly, Stone most comfortably sold timber, but fate had him presiding over the contentious efforts to strengthen wilderness protection. Given his priorities during reclassification, Northwest conservationists had no reason to expect much assistance from Stone in stopping in Kennecott.[28]

Yet a local newspaper quoted Stone in a way that sent conservationists scurrying with unexpected hope. To an audience that included Washington governor Daniel Evans and sixty legislators and other state officials, Stone said that an open-pit mine was not compatible with the wilderness character of the North Cascades. Conservationists seized the statement.[29] M. Brock Evans, a Seattle attorney whom the Sierra Club would appoint as its Northwest conservation representative the following month, wrote to Stone and praised him for taking this stand. (Evans added, encouragingly, that he had long hoped that conservation groups and the Forest Service might cooperate more.) Stone replied and, backtracking a bit, quoted long passages from the Wilderness Act establishing Kennecott's right to mine before sharing what was becoming the standard response: "Mining operations are not consistent with the maintenance of wilderness; yet we have the obligation to administer the provisions of the Wilderness Act, including the mining provision." The first clause gave conservationists like Evans hope, while the second swiped it away.[30]

While Evans flattered regional forester Stone, N3C's Goldsworthy went straight to the top, writing to Secretary of Agriculture Freeman (and copying another cabinet secretary, two U.S. senators, three U.S. representatives, a Supreme Court justice, and northwestern conservationists). Goldsworthy's appeal was multi-pronged but pointed to Stone's "not compatible" comment. The conservationist explained his views to Freeman. The Wilderness Act contained ambiguity by both allowing mining and demanding wilderness protection. But the law also contained avenues to resolve that ambiguity—in favor of wilderness, in Goldworthy's mind. Not only did the Wilderness Act allow the Forest Service to regulate mining, something Chriswell and others had already highlighted, it also allowed the government to purchase private lands held in designated wilderness. Further, Goldsworthy pointed out that the act required reports to Congress on the status of the National Wilderness Preservation System, the purpose of which, he interpreted for Freeman, was "to [apprise] Congress of the need for legislative action to assist the administering agency in carrying out [the act]." All these things—agency regulation, government purchase, status reports—showed Goldsworthy a path forward if Freeman would only step in that direction. Since Kennecott's mine was incompatible with wilderness, according to Stone, the Forest Service was required to inform Congress, Goldsworthy reasoned. He helpfully suggested to Freeman: "The Act appears to authorize you as Secretary to recommend that Kennecott's private lands be acquired, if you care to make such a recommendation." In case this strategy failed to move the agriculture secretary, Goldsworthy appealed to his bureaucratic pride: "If the Forest Service is, as it insists, the proper trustee for these public wilderness lands, and if the Forest Service cares about preserving these lands unimpaired, we think that the law provides it with a means to do it." The USFS looked beyond its purview if it weighed the economic import of Kennecott's copper; it should only consider the wilderness values of its lands. Goldsworthy's gambit leveraged Stone's "judgment" as the impetus to concoct a pro-wilderness, creative interpretation to resolve the law's founding ambiguity.[31]

Stone's public comments alerted more than local conservationists. After hearing of Stone's remarks, Representative Meeds wrote to Greeley, asking the associate chief to clarify Stone's statement. Were open-pit mines permitted in wilderness areas or were they incompatible with wilderness? The bureaucrat promptly assured the politician that the press had taken the forester's comments out of context. "This is one of the seeming riddles of the act," Greeley pointed out. The law required the agency to preserve wilderness except where the act provided otherwise. The agency could exercise some control of mining activity but not "nullify the law by imposing regulations that would make mining not possible." Greeley expected "a long and hot controversy over this" and chided conservationists for trying to reinterpret the act to regulate mining out of the wild.[32] Greeley's statement might have been the official agency line, but differing opinions appeared across the Northwest.

Mt. Baker supervisor Chriswell had become dogged and independent.[33] He met with Snohomish County officials and learned that Kennecott needed county permits for an open-pit mine. If that happened, county officials explained, the Forest Service would be called as an expert witness at a hearing. Informing the regional office of these developments, Chriswell used the opportunity to advance a strong position for the Forest Service, one that seemed inconsistent with Greeley's directions from D.C. and earlier instructions from Portland. Kennecott controlled eleven potential mines in wilderness areas across the nation. In some places, Chriswell explained, they could mine with few problems, but in other places "conflict with wilderness values will be so severe, as in this case, that our restrictions would make mining uneconomical except in a national emergency." This position differed little from previous utterances, but then Chriswell pushed further: "But we need to *establish our right* to determine this and to take the lead as the agency best qualified to protect these areas." To assert its right to regulate mining moved the Miners Ridge issue into national prominence and precedents, for it would establish—rightly, in Chriswell's mind—the agency's prerogatives and responsibilities. If the county planning commission held a hearing, the Forest Service, in Chriswell's view, needed to be unequivocal and state that Kennecott's open-pit plan was "completely incompatible" and that accordingly the agency would impose "tighter" restrictions than in "other areas where wilderness values [were] not as great." The USFS ought to use the hearing to educate the public about the "basic weakness" in the Wilderness Act, a weakness that hampered its raison d'être: to protect wilderness. Presumably this strategy aimed to build public support, perhaps even to call for legislative adjustments to bolster Forest Service power.[34]

Especially because they seem inconsistent with the agency's Washington and regional offices, how should we interpret Chriswell's ideas? In part, Chriswell sought to maintain bureaucratic prerogatives and power in a situation quickly leaving the agency's control. In part, he recognized a fundamental weakness to wilderness legislation and aimed to operate somewhat independently to advocate a stronger Forest Service position. And Chriswell was just getting started.

If winter was the season of meetings and correspondence, summertime meant field activity. And in summer 1967, forest supervisor Chriswell again showed initiative, taking a prominent group into the mountains on a show-me trip. Forest supervisors had done this before as an effective way to demonstrate conditions firsthand and generate support for agency initiatives. Sometimes the strategy backfired, such as when foresters in Oregon offered an informational tour in the Three Sisters Primitive Area in 1951 that galvanized wilderness activists against the agency.[35] This time Chriswell executed the trip with political and public relations precision by including members of the regional press and Seattle mayor Dorm Braman. A man with longtime experience in the lumber industry and a love for the outdoors forged as a Boy Scout leader, Braman was a lifelong Republican who readily worked across political lines.[36] Chriswell expertly used the opportunity to vaunt the area, explain the Forest Service's position, and break the news of agency restrictions on Kennecott's operation.

A beautiful, front-page, color photograph greeted *Seattle Post-Intelligencer* subscribers when they picked up the paper from their stoops on Sunday, August 27, 1967. In the background, a snow-covered peak rose up, its white blending into white sky so indistinctly as to make it almost impossible to tell where sky and summit met. In the middle ground, a ridgeline parallel to the distant Glacier Peak and covered in thick and dark green timber plunged toward the Suiattle River. The foreground, however, jumped with color. Bright green grasses, light and dark, were broken up by mountain wildflowers' yellows, purples, and reds. Spires from pine trees poked up from just over the nearest ridge, from which the photographer aimed his camera. This display of alpine glory, this wild tangle of beauty, was a "Stage Setting for an Open Pit Copper Mine," the photo's caption incongruously and provocatively put it. Turning inside to the accompanying story readers were greeted by more photos. Above the fold, covering almost half of the page, two black-and-white images contrasted life in the mountains, each with Glacier Peak rising in the background. In the left-hand photo, two men on horseback along a dusty trail were silhouetted by sunset, recalling the many generations of mountain travelers who relied on animal power. In the photo to the right stood a scaffold resembling a giant tripod holding a drill. A man standing on a makeshift platform halfway up suggested that the scaffold was probably forty feet high, providing a majestic view of Glacier Peak country. On the ground lumber, a ladder, and a mess of other indistinct equipment were strewn about. "Scratching the Back of Miner's Ridge," said the caption.[37]

Besides framing the controversy visually, the *P-I* reported the latest developments and agency perspectives. During the three-day pack trip that had preceded the front-page story, supervisor Chriswell had announced agency regulations, the most severe of which required Kennecott to dispose of mine waste so that it would "not affect stream flow or otherwise adversely affect land or water." According to the paper, Chriswell asserted that most Forest Service employees opposed Kennecott's mine "on general principles," noting the basic incompatibility of mining and wilderness. The Wilderness Act allowed what it termed "reasonable regulations" on mining, but Chriswell understood and explained the rub: "What we think is reasonable might not appear reasonable at all to the Kennecott people." Such statements, delivered within view of Glacier Peak, helped the Forest Service pitch itself as the responsible protectors of wilderness—an image that seemed a far cry from the truth to conservationists who had seen the agency exclude timbered valleys from wilderness areas. Kennecott's plans gave the Forest Service an opportunity to rehabilitate its regional image to conservationists, and Chriswell seized it.[38]

A feature in the glossy *Seattle* magazine followed later in the fall, in which Chriswell's summer tour group was referred to as "the North Cascades exploratory party," a rather grandiose characterization. The story, "Ride-In to Glacier Peak," outlined Kennecott's plan and expressed outrage and disgust. A Forest Service ranger, Calvin Dunnell, had briefed the party after a dinner of T-bone steaks and salad, not exactly rustic trail fare. Following the agency's developing script, Dunnell emphasized the Forest Service's regulatory demands, including close attention to controls on tailings, and shared what he saw as the best- and worst-case scenarios. The worst case would be that Forest Service regulations would "diminish the havoc Kennecott would wreak." The best case would be that restrictions would force corporate reconsideration and effectively prevent Kennecott from digging its pit. Dunnell's confidence that one or the other outcome would transpire reflected a common trait of agency staff, a sometimes overweening sense of their ability to rule their world, directly at odds with reality in the mid-1960s.[39]

The *Seattle* article described the Kennecott–Forest Service impasse over how to transfer ore off Miners Ridge, a specific point of contention that embodied the entire controversy. Kennecott wanted to use a conveyor, which would take ore down the steepest part of the ridge, and build a concentrator along Miners Creek, near where the "North Cascades exploratory party" had camped their first night. This site sat well within wilderness boundaries. Believing such a plan failed the "minimum impact" part of the Wilderness Act, the Forest Service devised an alternative plan to build a tunnel from the pit all the way down to the Suiattle River where it would connect to a spur rail line. It was a much more expensive yet less conspicuous option. Since agency and company reached no resolution that summer, Kennecott just drilled more tests and bided time.[40]

Besides the Seattle press, the *Everett Herald,* the largest newspaper in Snohomish County, sent a journalist along on Chriswell's summer trip. The lesson the *Herald* drew from Dunnell's briefings was that a court battle was inevitable. Either the Forest Service would issue strong regulations, which would bring a suit from the company, or the corporation would drill without permits, which would spur a suit from the agency. "The company apparently feels that our requests would increase their costs of operating to an unreasonable degree,"

Dunnell mused. "They have the right to take us to court to try to prove that the requests are unreasonable. We expect that they will." A legal contest might settle finally the strength of the Wilderness Act, the Forest Service's power, and mining companies' rights.[41]

Dunnell's statements, as represented in the press, drew Kennecott's attention and ire. Paul B. Jessup, the company's public relations director, wrote to Forest Service chief Edward Cliff, attaching the *Herald* article and inquiring whether Dunnell was "expressing the official position of the Forest Service or merely expressing his own ideas," adding that the ranger "appeared to be covering a great deal of territory and taking a lot for granted." Chief Cliff attempted to assure Jessup that there was no need for concern: "As you know, newspaper accounts are often not accurate reflections of what is actually said. Such a newspaper story does not, of course, represent official policy." Dunnell just meant that the Forest Service would deploy "reasonable regulations," as the Wilderness Act authorized. As with so many similar letters within the Forest Service, between the Forest Service and conservationists, and between the Forest Service and Kennecott Copper, the exchange ultimately clarified nothing and did not dislodge the impasse.[42]

Each man closed his letter to the other by remarking on the wilderness character of Miners Ridge. The tone was passive-aggressive in both cases. Jessup asserted that "so many people and pack trains" tramped the area that it was reminiscent "more of Times Square than a wilderness." The PR man seemed to be saying that the mine could not be incompatible with wilderness here because it was too crowded to have wilderness character any longer. Cliff acknowledged the longstanding popularity of the area but noted that "recent publicity . . . has increased public interest." The forester seemed to be responding, "You brought this on yourself."[43]

According to the *Seattle* magazine article about Chriswell's summer tour to Glacier Peak, the last night with Forest Service officials and journalists in camp "turned out to be nearly as spectacular as the day had been." After winding their way up countless switchbacks, they saw the site of the proposed mining operation. They also felt watched by Glacier Peak, always standing guard over the other mountains and valleys. They saw a tree-carving, famous among Northwest hikers, and camped in the ambience of Image Lake's iconic beauty. "The moon was all but full," the writer described, "and perhaps under its influence, the horses, grazing nearby, ran amok, frisking and whinnying and almost trampling the campers who had shunned the stuffy confines of a tent." This image, of horses running free under the wild moon, symbolized what might be lost. And why some might fight for it.[44]

The Forest Service occupied an unenviable position as Kennecott began pressing forward. The agency managed wilderness across nearly half a million acres surrounding Glacier Peak. While preserving and protecting that land, it sometimes differed with local conservationists, favoring timber values in some places more than wilderness. Nevertheless, the agency felt no loyalty to hardrock mining. When Congress wrote the Wilderness Act and compromised to allow mining and prospecting, it also forced ambiguity onto the agency: an unequivocal right for miners to mine and a series of clauses allowing the Forest Service to regulate such activity.[45] The agency struggled to articulate how to respond to Kennecott's activity high up in the Cascades. This uncertainty was compounded by the decentralized nature of the Forest Service, with district rangers and forest supervisors working locally while regional foresters tried to carry out their priorities in response to the Washington office, which answered finally to the secretary of agriculture, who soon would weigh in publicly and powerfully. Amid all those bureaucratic layers, no one could be surprised when priorities or messages were mixed.

As the dance between the agency and corporation quickened, agency officials explored ways to move outside the normal routine, to improvise new steps. Knowing full well it could not prohibit mining, the agency suggested a regulatory arsenal that complicated Kennecott's pursuit of profits. And if regulations were insufficient to deter Kennecott, which seemed confident in its rights, then the Forest Service would begin mobilizing a public campaign to discredit Kennecott's plans. Doing so would help to bolster—or rehabilitate, depending on one's perspective—the agency's reputation as stalwart wilderness protector. The Forest Service did not face Kennecott alone, of course. While foresters faced corporations across tables, in newspapers, and through private correspondence, conservationists prepared to meet the corporation face to face.

FOUR | The Summit

SUMMITS AND WATERSHEDS. BOTH are features of landscapes. Both have dual—and contradictory—meanings. Both function as useful, if easily mixed, metaphors. Watersheds are areas where waters gather as an integrated unit, such as the Suiattle River watershed, a defining feature in a broader landscape. But watersheds also denote a break, a divide between one unit and another. When we cross a watershed, we are moving from one condition to another. Just so are summits. They are the places where all sides—slopes on a mountain—come together in a singular spot. They mark the place where we can see all sides. Heads of states are said to be at a summit when they meet to discuss weighty matters, such as nuclear disarmament. So "summit" is a useful metaphor to describe the January 1967 San Francisco meeting of six men—four from conservation groups, two from Kennecott Copper Corporation—that gave all a view of the surrounding terrain. For the first time, the company formally shared its detailed mining plans, and conservationists countered with their specific objections. Rather than reaching a disarmament deal at this summit, a crevasse between them widened and made future navigation trickier. At the summit, conservationists crossed a watershed.

Just like any ascent to a mountaintop, getting there was indirect, and traversing the ridges oftentimes required delicate care and deliberate choices. After learning that Kennecott representatives had met with local Forest Service administrators but before the mining company announced its plans publicly, the president of the Sierra Club, George Marshall, wrote individually to members of Kennecott's board of directors as well as to the corporation's president, Frank R. Milliken, to whom he also gave a copy of the Sierra Club book *The Wild Cascades.* Patrick Goldsworthy, president of the North Cascades Conservation Council, drafted the letters for Marshall, hoping they might "raise a reasonable doubt in the minds of one or two" of the directors. In particular, Goldsworthy and Marshall hoped to persuade the board that a large public fight might "take some of the lustre off their image" and not be worth it for the comparatively small productivity the mine promised. They posed the question pointedly: "Has your management weighed the marginal nature of the profits which might accrue to Kennecott from this operation, compared to its other far larger operations, against the possible loss in public good will that could result from damaging scenery which qualifies for and is actively being considered for a national park?" This, then, was the initial approach conservationists settled on: moral suasion, directed at the company in the public realm, where conservationists hoped it might attract attention.[1]

As he lobbied the board, Marshall bolstered the Sierra Club's case in two ways. First, he included several passages from national agencies and federal studies about the "superlative alpine scenery" to show that multiple groups—government entities, no less—found the North Cascades worthy of protection. Second, his letter emphasized the Sierra Club's long interest in protecting the Glacier Peak area, its recent successes in helping to block dams in Grand Canyon, and the imminent success in supporting the proposed Redwood National Park, which included getting a logging moratorium in place while the conservation politics sorted out. In other words, Marshall put Kennecott on notice that conservationists would put up a formidable challenge, the sort they recently had been winning.[2]

While Marshall plowed ahead, Kennecott's upper management sought a conciliatory path. Responses from members of the board followed a pattern, likely because they had discussed it at their most recent meeting. Individuals typically expressed a personal interest in conservation and scenic beauty, sometimes even referencing time spent in high mountain ranges such as the North Cascades. Then they pivoted to the national need for copper to fulfill strategic needs—national security *and* corporate financial health. They hoped "something mutually constructive and satisfactory" might come of the scheduled meeting between Kennecott and conservation representatives. This accommodation was possible, they believed, because, in their view, operating an open-pit mine could "maintain the general area" without harming scenic qualities (an incredible statement for anyone who had seen Kennecott's massive open-pit mines in Utah or New Mexico). Despite the incompatible outlooks, individual directors believed in finding common ground.[3]

President Milliken also responded in an accommodating tone and revealed personal history. Like the board members, Milliken hoped for a peaceful resolution that met the needs of the club, the company, and the nation. Milliken had managed mines in wilderness settings, having directed a titanium mine in the 1940s in New York's Adirondacks. The mining man assured the conservationists that despite concerns at the time, "we were able to develop procedures which minimized the impact on the area and yet at the same time made it possible to recover a natural resource which was much needed by the United States in the war period and is still needed today." The message was clear: a time of war required mineral resources, and Milliken was just the man to ensure minimal harm to nature. The corporate president also appreciated the book and informed Marshall that he had already enjoyed others in the Sierra Club's series. He even ordered three more copies of *The Wild Cascades* as well as other Sierra Club large-format books. A note on the order—made by the company's public relations director—stated: "Because of their great significance in the literature on the subjects covered, Mr. Milliken thought we should have these in our N.Y. Office library." *The Wild Cascades,* with its photos of beautiful North Cascades wilderness, might then have graced shelves in a boardroom where men agreed to befoul the very slopes it pictured.[4]

This initial correspondence found only superficial common ground. Everyone likes stunning scenery, after all, so conservationists, the board, and the president agreed on that. And the correspondence included democracy's favorite shibboleth about working for a resolution that served everyone's interests. Taking them at their word, we can imagine high hopes as they planned to meet in San Francisco. The Sierra Club and its regional counterpart, the North Cascades Conservation Council, might have felt emboldened by the strength they had marshaled to prevent dam building in Dinosaur National Monument and Grand Canyon, as well as the support they had generated for the Wilderness Act and for what seemed to be an imminent national park for the redwood forests of California. Yet the mining industry was different. Exceptions to allow mining in wilderness were compromises that conservationists had accepted to get the Wilderness Act through Congress. Minerals necessary for modernity's monuments and war's weapons came from pits. The government's policies and laws promoted mining as the highest use on public lands, and incentives through the decades had favored extraction. When conservationists sat down across from a corporate vice president and public relations specialist, Kennecott's long history and their privileged position vis-à-vis federal policy sat with them.

Copper as a metal and Kennecott Copper Corporation as a business symbolized the modern industrial order. As much as anything, electricity represented modernity, and copper constituted the central element necessary for wiring the world and moving it into a second industrial age. Billed as a clean alternative energy, electricity promised an industrialism that improved on the sooty cities powered by coal and characterized by steel manufacturing.[5] The United States led the way in electrification, consuming roughly 50 percent of the world's copper supply by the 1890s.[6] The rising and insatiable demand for copper pushed American mining companies toward locating and then depleting known supplies, ore bodies with high concentrations of copper. Discovering high-grade ore in Alaska's Bonanza mine in 1897 and developing a new method for open-pit mining in Utah's Bingham Canyon in 1904 fueled the American copper industry's expansion. These two operations combined with a third, the Chilean mine El Teniente, to comprise the multinational Kennecott Copper Corporation, organized in 1915. Kennecott, in its international scope and its deployment of innovative technology, epitomized modern industrial copper production.[7]

As copper became central to this new industrial age, Kennecott and other copper producers quickly faced a dilemma. High-grade ore deposits had been found and depleted in short order. To continue the electric revolution meant figuring out ways to recover copper out of low-grade ore bodies profitably. Geologists knew, for example, of the massive copper quantities at Bingham Canyon, outside of Salt Lake City, but the copper being mined there amounted to a concentration of roughly 2 percent, and that was declining rapidly.[8] The costs of extracting ore in traditional underground mines exceeded the returns. A mining engineer from Missouri, Daniel Cowan Jackling, calculated that a company could speed up the extraction rate using powerful machines working in a massive open pit. A technological knot entangled all this. As it was developing in the United States, modernity demanded electricity. Electricity required copper. Copper mining required power. The Bingham Pit, as it soon came to be known, used as much electricity as a city of one hundred thousand people. Mining in open pits, as historian Timothy J. LeCain put it, only "worked in the era of powerful but crude engines, energy-hungry steam, electric, or diesel monsters that could do the work of thousands of men or animals in a fraction of the time." Open-pit mining constituted "mass destruction," in LeCain's apt phrasing, "a technological system for cheaply extracting huge amounts of essential industrial minerals from the earth's crust." Such mines destroyed mountains by systematically dismantling them. Mass destruction was yoked to rising mass production and mass consumption in the United States, although the links were distant, complex, and largely kept hidden from public view. The first dynamite exploded in Bingham Canyon in 1906, and in less than five decades more than 60 percent of the world's copper came from open-pit mines, a destructive revolution in mining processes. Open pits proliferating around the globe meant that twentieth-century electrical gadgets would continue to run and that more would be manufactured, spurring on the need for more mines.[9]

Even though the new economy of scale brought by open-pit mining increased corporate profits, the copper industry could not avoid the volatility of global supply and demand fluctuations.[10] These economic cycles were tied to new ore discoveries and rising demand during wartime or periods of technological innovation. While it may have seemed to some as though an invisible hand guided the ups and downs of mineral prices and markets, corporate and government policies worked as visible hands that shaped copper's position in the U.S. political economy. Following World War I, copper demand and prices declined, and corporations compensated by expanding the industry internationally. Copper sources also declined after locations in Michigan, Arizona, and Montana had been mined for decades, and as demand increased again, American investors looked abroad for new sources. Companies like Kennecott emphasized development overseas and sometimes shortchanged domestic investment, showing that international development was not only about finding copper but also about increasing profitability. As a result of this trend, U.S. copper sources comprised an increasingly smaller share of global copper production. In the first decades of the twentieth century, American copper constituted up to 60 percent of the global share, but by 1960 it did not even capture a quarter of it. At the same time, global production tripled. This shift toward decline in American copper production was not linear—constant fluctuations characterized it—but the long-term trend toward permanent decline could not be masked by the mid-1960s.[11]

Through the mid-twentieth century, the copper industry's fortunes remained entwined with a series of global events and government policies, creating the context that shaped Kennecott's plans for Glacier Peak. The pre–World War II high for U.S. copper consumption came in 1929, when it topped three million tons. (After the war, only once would consumption *not* reach that mark, showing a seemingly inevitable rise associated with a growing American economy.) The Great Depression then devastated the industry, with copper prices dropping in 1932 to six cents per pound. World War II prompted rising consumption and prices, as war efforts required full production even with short supplies and personnel. Federal policies at the time directed much of the copper industry's business through price controls, subsidies, loans, and other measures. Corporations like Kennecott left the war weaker because of these policies, but after restrictive government policies expired, postwar prices rose from thirteen cents a pound to twenty-three cents per pound, pushing companies toward higher profit margins.[12]

Nevertheless, the federal government remained concerned about copper supply in the postwar era and continued to have a hand in the industry. Fearing shortages, the government sought to create strategic reserves, but congressional budget-pinching constrained those efforts. Harry S. Truman created the President's Materials Policy Commission to recommend long-term solutions, and its five-volume report was filled with pessimistic predictions that consumption would outpace domestic production. In 1950, shortly after the start of the Korean War, Congress responded to this well-defined and -publicized fear of shortages with the Defense Production Act, intended to promote production. The legislation offered various incentives to industry: accelerated depreciation to improve corporate tax situations, favorable loans through the Reconstruction Finance Corporation, and long-term purchase agreements. These terms sparked the most serious investment in domestic production between World War II and the Vietnam War. While U.S. copper companies stabilized somewhat between the Korean War and Vietnam, businesses responded not by reinvesting in its

U.S. production system; instead Kennecott, for instance, paid out 86 percent of its profits as dividends to investors.[13] These government and corporate strategies weakened the copper industry's strong foundations of half a century before and focused on short-term profits instead. Thus, Kennecott stood in a precarious position at the dawn of the Vietnam War. Its domestic production had stagnated; its profits remained substantial but were declining; its prospects seemed comparatively bleak. New sources of copper would have seemed an attractive investment.

That is not to say that Kennecott only struggled or was weak. The company enjoyed a global reach, stretching from Canada to Chile to Africa and Australia. Although its Alaska mine closed in the 1930s, others continued producing, and by the 1960s the company had expanded further. Properties throughout the West—in Arizona, Nevada, New Mexico, and Utah most prominently—joined productive mines in Canada and Nigeria as well as exploration in Australia and Puerto Rico. In 1965, just before announcing its Glacier Peak project, Kennecott mined 619,868 tons of copper, 451,645 tons of which came from its western United States division, so although it was a multinational company, it remained rooted in the West. Its investments allowed the company to become cash-rich, claiming more than $711 million in accounts at the end of 1967.[14] With copper mining in the United States in steady decline, Kennecott remained an impressive corporate force, but the company could not rest on its laurels; it needed continued profits.

Meanwhile, as in the past, war prompted federal intervention in the copper economy. The General Services Administration had been formed in 1949 to improve federal administration and management of resources including stockpiling strategic materials as part of President Truman's general efforts to support preparedness and security.[15] In the spring of 1966, the GSA announced a new incentive, the Copper Production Expansion Program, to bolster the war effort in Southeast Asia by spurring American copper production. The GSA specifically requested of corporations that new mines be "brought into production early" by promising "prompt consideration . . . to such expansion proposals." Besides providing purchasing contracts, the GSA assured assistance in acquiring equipment for companies struggling to obtain scarce supplies. Business loans, technical assistance, and public works also were options for this program. By August around one hundred firms had expressed interest, and twenty filed required plans.[16]

Kennecott was among those interested. The corporation shared with the GSA its plans to bring its Glacier Peak property into production by 1970. Kennecott did not request a federal subsidy, even though it expected high operating costs. Instead the company wanted help from the Forest Service and Department of the Interior to expedite processes so the company could acquire land next to the mine to dispose waste and tailings and for access roads and power rights-of-way.[17] Additionally Kennecott recognized the potential challenges that wilderness status might entail and sought assurance that federal agencies would not place "prohibitive" restrictions on the company and make the enterprise uneconomical. In other words, Kennecott did not seek the GSA incentives; it merely desired federal grease on bureaucracy's wheels.[18]

In the end, Kennecott never formally applied for the program. Indeed, officials from the GSA and the Department of Defense eventually informed Scoop Jackson that they likely would have denied Kennecott's application on both practical and what might be termed aesthetic grounds. The GSA intended that the expansion program increase domestic copper production rapidly, and Kennecott's estimate that it would take more than three years to get up and running seemed slow. More importantly, an assistant secretary of defense confirmed a GSA commissioner's assessment that any copper supplied from Kennecott's efforts would have been insufficient "to outweigh other important considerations, such as the inevitable damage to the natural beauty of the wilderness area." According to another official, the GSA was "aware that the property was in a Wilderness and the agency did not want to be involved in such a situation." Officials in federal agencies other than the Forest Service recognized the fraught political—if not, legal—position that Kennecott was pursuing at Miners Ridge.[19]

Nonetheless, the GSA program gave Kennecott important public cover by emphasizing national wartime needs, and the corporation used this against its opponents. When Kennecott initially announced its intentions, the company explicitly referenced wartime copper needs. The company's new mines division head, M. J. O'Shaughnessy, explained that the government was requesting more copper to combat a global shortage: "Our government is hoping that the mining industry will come up with more sources of copper in this country. We, therefore, must look at all sources and plan ahead."[20] Members of Kennecott's board of directors cited their "patriotic duty" to feed more copper into the nation's military-industrial complex. Conservationists had to work against those forces. They knew that Miners Ridge contained minerals, but they believed its low-grade ore, protected status, and isolation would make exploitation "uneconomical" for any firm interested in dismantling the mountain. But the war in Vietnam altered the metal market to favor mining companies.[21] The very existence of the GSA program—even though Kennecott never participated—highlighted the improved market conditions and a federally sanctioned effort to find and exploit copper deposits. With a needy government and a powerful firm, the circumstances favored Kennecott, and odds as steep as North Cascades slopes confronted conservationists when they finally met with company representatives.

On January 30, 1967, two Kennecott officials and four conservationists met in San Francisco and over the course of three hours shared plans and objections. This was the first opportunity to meet face to face since Kennecott's announcement of its intent to mine in Glacier Peak Wilderness Area. From the corporation came the vice president in charge of mining, C. D. Michaelson, and a corporate relations representative. The conservationists were Sierra Club president George Marshall, North Cascades Conservation Council president Patrick D. Goldsworthy, Sierra Club conservation director Michael McCloskey, and Sierra Club Northwest conservation representative Rodger Pegues. Perhaps the roster reflected an advantage for wilderness advocates. Not only did they outnumber the mining executives and combine organizational forces, they were also all closer to home in the city that held the Sierra Club's offices. Whether this show of strength fazed Kennecott's representatives is lost to the historical record. However, the topic and tenor of their discussion established the terms for the forthcoming campaign.[22]

The Kennecott executives presented and justified their plans. The Miners Ridge operation would amount to about 7 or 8 percent of Kennecott's total copper production, perhaps twelve thousand to fifteen thousand tons of copper, which was less than 1 percent of the nation's total production and less than half a percent of Americans' copper consumption. The company anticipated constructing a road to Miners Ridge "as soon as the snows are gone"—a phrase that hung over conservationists as a threat every winter—and believed the mine would produce for twenty or thirty years and employ two hundred locals. Two-thirds of the excavation would be waste rock, while the ore's low grade (1 percent) required a concentrating mill using a flotation process to get it to 36 percent. Although most of the conservationists' attention focused on the open pit, the processing generated heavy extraction costs. The flotation process would require five thousand tons of water to process five thousand tons of ore daily. Waste would cover two hundred acres with tailings. Electricity demands from the nearest source, in Darrington, would reach up to seven thousand horsepower. This endeavor represented significant industrial intrusion beyond a simple hole in the ground, and it joined larger Northwest enterprises represented by proliferating hydroelectric dams and growing manufacturing giants like Boeing and Weyerhaeuser.[23]

Kennecott justified this intervention in detail. In fact, the reasoning seemed almost as developed as the mine's technical plans, suggesting that company officials recognized precisely how controversial a wilderness open-pit mine would be. Michaelson explained three needs the Miners Ridge mine would help meet, and he framed each as contributing to national priorities without acknowledging corporate benefits. First, the Department of Defense had called to increase production to mitigate copper shortages and offered subsidies to develop mines that had been "kept on the back burner." Next, the United States was a net importer of copper, meaning that the balance of payments did not favor the nation—a sign of economic vulnerability in international markets. Kennecott understood from government contacts that this imbalance would last through the next decade but new mine production could improve the disadvantage. Finally, copper stockpiles had grown dangerously low, amounting to one hundred thousand tons instead of the typical seven hundred thousand tons, a level that Kennecott officials believed was still too low. Michaelson characterized this situation as disgraceful. In short, he said that the nation was calling for Kennecott to help right the ship.[24]

Michaelson also expounded on the broader circumstances that required more copper mines. Per capita copper consumption in the United States tracked at roughly 17.5 pounds per year. To keep up with the rising U.S. population and its standard of living, Michaelson reasoned that more copper needed to be mined. He explained that it was "boom times for the copper companies all over the world." Production kept expanding. Three daily shifts, seven days a week, was the order of the day. But older mines inevitably produced less and less as ore bodies were depleted, so to keep pace copper companies needed to open seven or eight new low-capacity mines for every older high-capacity mine that bottomed out. This new reality required copper corporations to search near and far. Michaelson told the conservationists in San Francisco, "We must exploit all the remaining possibilities we have and we are looking all over the world." To continue to be profitable—not to mention serving national defense needs and satisfying consumer quality of life demands—the vice president of mining indicated, the company would be irresponsible if it passed over Glacier Peak.[25]

So Kennecott Copper Corporation planned the mine and assured the gathered conservationists that in long run wilderness would return to Miners Ridge. But the company's story failed to persuade the activists or allay their fears. When Kennecott explained it would plant Australian vetch on the tailings and dump piles, conservationists countered that introducing exotic species violated the Wilderness Act. When the company claimed that a few avalanches and slides would eventually wipe out the road and soon make it unnoticeable, they scoffed. When Michaelson assured them that the tailings dump would remain "behind bulkheads," with no way to pollute streams, conservationists reminded him that a few miles away the Holden Mine failed to keep pollution from Railroad Creek using the same bulkhead system. A letter soon after the summit from Kennecott's director of corporation relations reinforced the way the company minimized risks and problems through misdirection by citing how the open pit would eventually become "an attractive lake."[26] In contrast, conservationists like Goldsworthy believed the wilderness would "be shattered" by Kennecott's planned activities.[27]

At the summit, which brought the sides together in the same room, conservationists viewed gaping chasms between them and left committed to stopping Kennecott's actions. As protectors of wild places, they felt horrified at the corporation's pledge to dismantle a mountain and shocked by the company's inability to recognize the fundamental differences between an alpine lake and a filled-in pit, a talus slope and a tailings pile, and a road and a trail. Kennecott may have shown its clear intentions, but Goldsworthy maintained that conservationists had expressed "with equal clarity" that they had "*not* accepted this plan as being inevitable."[28]

As Goldsworthy groped toward making meaning of Kennecott's planned assault, a strategy emerged—one that foreshadowed much of the ensuing campaign. First and foremost, Goldsworthy framed the story by focusing on Kennecott as a greedy villain, announcing that "[t]he corporate ruthlessness of this company must be exposed to the entire nation." This exposure had to drive public, grassroots protest to reach the president and members of Congress, which could prompt political action. Goldsworthy could not have anticipated the many unique and powerful ways the public would highlight what they interpreted as corporate greed, but that theme remained steady throughout the campaign. This strategy offered a political and moral approach, not a legal one. That is to say, it meant to persuade the public, the corporation, and policymakers that the proposal was untenable, immoral, and outside the public interest—not that it was illegal. Second, and more narrowly strategic, conservationists pinned hope on incorporating Miners Ridge into a national park. An ongoing campaign to create a North Cascades National Park gained steam right as Kennecott upped its exploration activity. Although wilderness protection primarily stayed within the Forest Service, the National Park Service (NPS) enjoyed stronger ability to stop mining. So, while Goldsworthy and his colleagues worked to encourage the USFS to regulate and control the wilderness in all allowable ways, they pursued NPS protection as part of a larger regional program that could stop mining. Third, and broadly strategic, conservationists would cooperate. Goldsworthy explained to North Cascades Conservation Council members that the Sierra Club was turning its attention to the North Cascades. In fact, the Sierra Club had just defeated efforts to dam the Colorado River in the Grand Canyon and now turned to the Redwood National Park and North Cascades as their top two national priorities.[29] The Northwest would not be a rearguard action but a prominent front in the national conservation campaign. Central to that strategy would be coordinating efforts between the Sierra Club and numerous northwestern organizations, demonstrating the collective effort and high priority the region occupied nationally. Last, they floated a few specific ideas that seemed exploratory in nature. These ideas included a five-year moratorium on developing the mine to give Congress a chance to enact a national park and to encourage Kennecott to develop three of its other projects that were not located in wilderness areas. One fundamental principle also undergirded conservationists' resolve: Kennecott could not be allowed to even begin the project. On this, Goldsworthy and the others were emphatic, and it set the tone for the long trek ahead.[30]

After the summit, Sierra Club president George Marshall wrote to Frank Milliken, president of Kennecott, purposely going to the top and not the vice president he had just met. Steadily polite, Marshall let the Milliken know that conservationists "were far from reassured" by plans and promises offered at the meeting. "The danger to this area seems greater than ever," he wrote. Marshall explained that Kennecott had failed to persuade the conservationists that the operation could avoid "serious damage," improve the balance of payments, or reduce the need to import. In short, none of Kennecott's arguments held water, much like conservationists imagined the company's leaky tailing ponds behind bulkheads, befouling mountain streams. The uniqueness of the Glacier Peak wilderness—characterized by the NPS as "outranking in natural beauty any existing or proposed national park in the entire continental United States"—made the mine fundamentally incompatible, an inappropriate presence within wild lands. The comparatively small amount of ore would play an insignificant role in changing balance of payments shortfalls. Marshall put Milliken on notice. The Sierra Club and a group of coordinated conservationists would push for a national park. In the meantime, they sought a five-year moratorium to give the public a chance to purchase Kennecott's claims—a persistent idea that floated through these years. Since the copper stores were so small, "its use or its eventual non-use will have no measurable impact on our economy."[31]

Goldsworthy wrote Kennecott's president the same day as Marshall, with much the same message, although he emphasized more strongly how the mine would contravene the public interest. Goldsworthy made explicit what Marshall implied. Then he appealed to corporate logic. He imagined that "permanent scars left on the land" would forever mar Glacier Peak, but he also told Milliken that if he pursued the mine, "a permanent tarnish [would be] left on the corporate image of the Kennecott Copper Corporation." In closing, Goldsworthy told Milliken that the company's planned mine near Glacier Peak would be a national issue with a national campaign against it and that publicity would show Kennecott in poor light. That he promised.[32]

As with many historic summits, the one in San Francisco failed to resolve differences, but at least it clarified, for each side, their commitments. The battle now was officially joined. As conservationists searched for ways to publicize the egregious action Kennecott contemplated, they found allies in high places who helped them make a case for the public interest.

FIVE | The Secretary

SECRETARY OF AGRICULTURE ORVILLE Freeman faced a challenging task and uncertain audience, but he seized the opportunity. He had been invited to address the Sierra Club's biennial wilderness conference: fifteen hundred people gathered at the San Francisco Hilton in the spring of 1967 to share ideas about "wilderness and the quality of life." Freeman's department included the Forest Service, the agency that controlled the most acres of protected wilderness and the one most beleaguered by critics. Its long-standing multiple-use policy of production rubbed uneasily against the Wilderness Act policy of protection, and the agency's track record in the Pacific Northwest showed an almost-eager readiness to shrink wilderness areas more than secure them.[1] Freeman needed to prove to his audience that he and the Forest Service offered the best option for protecting wilderness.

Freeman chose to focus on the North Cascades to make his claims—a calculated, but risky, choice. The Sierra Club, which the *San Francisco Chronicle* then characterized as a "militant conservation organization," called for the Forest Service to surrender its North Cascades holdings so the National Park Service could administer a North Cascades National Park.[2] Losing land to national parks always rankled the Forest Service, and the major parks in Washington State—Mt. Rainier and Olympic—had been stolen from the agency.[3] To prevent a third park in the state from being poached from USFS lands, Freeman wanted to prove that the Forest Service could offer strong safeguards because more wilderness contests would arise across the country and because the North Cascades was heating up as a major national issue. Kennecott Copper Corporation lent Freeman a perfect foil that allowed him to talk about the area without even mentioning the proposed park. By targeting a corporation, Freeman could sidestep—at least for a time—the intra-governmental rivalry plaguing him to hit back hard against excessive commercialism and to praise wilderness. As Freeman rose to take the podium on the evening of April 7, he mustered all his eloquence for the cause of maintaining wilderness.

The Sierra Club's biennial wilderness conferences, begun in 1949, played a significant role in advancing the nation's conversation about wilderness. In the mid-twentieth century, the Sierra Club itself was transitioning from a regional club based in California, focused on protecting parks and promoting outings, to a national organization with broader political concerns. These conferences helped advance this evolution.[4]

First proposed by Norman Livermore, the wilderness conferences addressed an emerging and pressing backcountry issue. A boom in outdoor recreation, aided by Sierra Club outings that brought families into the mountains and inculcated a love of the outdoors, was evidenced in a sevenfold increase in use of Forest Service primitive areas between 1941 and 1952.[5] This recreation wave threatened to erode the sanctity of the backcountry. Livermore's idea for the wilderness conference came from direct personal experience. He worked summers packing for Sierra Club trips and recalled using more than 120 horses and mules to carry the materials needed for two hundred campers.[6] Such a large crew was bound to leave an impact on the land, and Sierra Club leaders hoped the wilderness conferences might address this problem and devise a set of rules and ethics for treating wild areas.[7] David Brower, the club's executive director, once explained that the conferences brought people together to hear "how to enjoy wilderness without wearing it out, or, stated another way, how not to love it to death."[8] However, Brower and others soon recognized that the threat wilderness lovers posed to the land paled compared to threats from those who wanted to remove or shrink wilderness boundaries or who wanted to make exceptions to strict regulations.

So the conferences evolved, becoming more political and strategic, just as the club itself evolved.[9] They became occasions for twenty-five years during which people shared and honed ideas of wilderness as a place, philosophy, and policy. Although the founders of the kindred Wilderness Society in the 1930s had articulated key ideas about wilderness within broader intellectual currents, their ideas received comparatively scant attention or development in the subsequent decade.[10] The Sierra Club wilderness conferences helped revive this intellectual tradition and added a political element, as the conservation community wrestled with the place of wilderness in America's larger social and political complex.

Thinking of wilderness in policy terms represented something of a shift, something that had until then been largely discussed within the Forest Service. Sierra Club organizers invited other conservation clubs, federal land managers, resource user groups, and concessionaires to attend the conferences, thereby guaranteeing a cacophonous lack of consensus. But that lack of uniformity meant greater energy and perhaps faster evolution of ideas. Longtime Sierra Club leader

Michael McCloskey recalled the clever strategy of inviting some speakers who had little experience with wilderness or expressed ambivalence about it. The goal was to get them to listen rather than speak.[11] And the ideas flowed and rushed forward. In no time a federal wilderness bill became a central focus, and then the conferences built support for what would become the Wilderness Act.[12] For example, Howard Zahniser, the executive secretary of the Wilderness Society, used the 1951 wilderness conference to first broach the idea of a national wilderness system protected by Congress. So these conferences presented opportunities for wilderness advocates to share and germinate ideas as well as to present them to a broader public, especially because the Sierra Club published proceedings.[13] Finally, the conferences also included notable participants over the years from government, including Supreme Court justice William O. Douglas and Senator Scoop Jackson, both of whom were from Washington State and would play roles in the North Cascades.[14] By the time Secretary Freeman took to the dais in 1967, the Sierra Club's wilderness conference had become an important and growing institution among conservationists.

Freeman occupies an odd place in conservation history. He brought little relevant background yet sat in a central position as wilderness debates unfolded in the 1960s. A three-term Minnesota governor from 1954 to 1960, Freeman had been tapped by President John F. Kennedy to head the Department of Agriculture, an appointment he kept through Lyndon B. Johnson's presidency. He was born in Minneapolis in a state with little federal land. Freeman became close friends with Hubert Humphrey, who later first introduced the wilderness bill in the Senate.[15] They remained longtime political allies, and Humphrey had first suggested Freeman for the cabinet post. As secretary of agriculture during President Johnson's War on Poverty, Freeman presided over the food stamp and school breakfast programs while also pushing to expand U.S. agricultural exports.[16] None of this suggested an interest in forestry or wilderness. When the Senate Committee on Agriculture and Forestry held confirmation hearings for Freeman, it largely overlooked the Forest Service. Only Senator Maurine B. Neuberger of Oregon asked for assurances, saying, "I hope that you are interested in forest conservation." She and Freeman spoke specifically, and briefly, only about reforestation and tree farms. Wilderness advocates were not likely filled with great hope.[17]

Despite having no track record in natural resource matters, Freeman presided over the Forest Service, the only agency at the time with designated wilderness. Since the Kennedy administration supported the wilderness bill, Freeman and Secretary of the Interior Stewart Udall faithfully aided its legislative path, a marked change from the two departments during the previous administration.[18] Freeman also showed signs of being persuadable. As a young Sierra Club field organizer in the Northwest, McCloskey learned of logging plans in several wilderness high country areas and enlisted two powerhouse senators from the Northwest—Wayne Morse of Oregon and Scoop Jackson of Washington—to investigate. Their pressure led to Secretary Freeman issuing a moratorium on logging in high country. Although the moratorium was vague and reversible, it was something. More important, it represented the first time local activists succeeded in getting congressional pressure to bear on the Forest Service.[19] On plenty of other occasions Freeman favored timber cutting and did wilderness no favors. He helped perpetuate the so-called purity doctrine that the Forest Service used to forestall wilderness designations; when any signs of historical development were evident, the agency argued that wilderness could not be declared.[20] Freeman's ambiguity was fitting since the Forest Service itself represented ambivalence, not least in its history with Glacier Peak.

So, when he was given the stage at the Sierra Club's conference, it was not known what Secretary Freeman would say. Would he promote USFS wilderness policy or its multiple-use mission? Would he criticize the Wilderness Act for tying his hands, or would he express reluctant accommodation in a system of laws? Would he speak as a politician-turned-administrator and tout programmatic priorities or as an impassioned conservationist who appreciated the irreplaceable values wilderness provided?

Much of Freeman's address fit a politician's typical banquet speech. (Every talk Freeman gave was remarkable since a World War II jaw wound had required him to relearn how to speak.[21]) The secretary presented his personal wilderness bona fides by talking about where he had been (e.g., Montana's Bob Marshall Wilderness, Idaho's Sawtooth country, Washington's North Cascades, and Minnesota's Boundary Waters) and who he had accompanied, including other presenters at wilderness conferences past and present (e.g., Forest Service chief Edward Cliff and Supreme Court justice William O. Douglas, who sent greetings to his friends through Freeman). Freeman praised the Department of Agriculture's accomplishments over decades of conservation work, highlighting repair work done by the Soil Conservation Service. He ended his address with a prospectus for the department, explaining that the USDA would be investing more money, improving land use policy, and concentrating on preventive conservation as it prepared for the twenty-first century. None of this qualified as unusual.[22]

But Freeman had attended the wilderness conference to talk about wilderness, and the middle of his address—its heart—hit the theme hard and focused pointedly on Glacier Peak. He waxed poetic: "The Cascades are an ocean of mountains, frozen in space and time, wave after cresting wave of stone, dotted with the deep blue-green of alpine lakes, laced with the glacial remnants of another age. . . . To call them 'America's Alps' understates the case. They are uniquely American, and if Americans destroy their character we will not see their like again." Sparing no rhetorical flourish and sounding more like John

Muir than a cabinet secretary, the Minnesotan explained to his audience that to understand the North Cascades, one needed to use one's senses: "to hear the wind above timberline, a voice like all the rivers in the world flowing over a thousand miles of granite and green; smell the pine; feel a pebble polished by eons of time."[23]

Freeman relied on two enduring tropes in the American environmental tradition: timelessness and American exceptionalism.[24] Wilderness, of course, is not timeless but historically grounded like everything else, and the very fact that Freeman faced this audience with the North Cascades dilemma reflected that.[25] By invoking the *essential* nature of American wilderness, Freeman set up two appeals. One nodded to a patriotic or nationalist argument that frontier wilderness made Americans, an argument that intellectuals and policymakers—Frederick Jackson Turner and Theodore Roosevelt most prominently—had been making for nearly a century.[26] The other pointed toward the irreplaceability of such places, a reminder that not all resources—even under the Department of Agriculture's jurisdiction—were renewable and so required strict protection. Freeman's words transported people from a hotel banquet room to a beautiful alpine scene with powerful sensory imagery. He also rooted it effectively in both the specific place of Miners Ridge and a general theme of the enduring values of wilderness against transitory human desires.[27]

Then Secretary Freeman challenged Kennecott Copper Corporation directly. He approached his critique in spiritual, legal, and political terms—each of which strategically advanced the argument against mining near Glacier Peak. Freeman, a Lutheran deacon, wondered if the copper had been placed at Miners Ridge "by a wise Creator to test whether man could forego material riches for the fullness of the spirit."[28] The point recalled generations of Americans who found in nature something divine, something beyond Mammon. Muir, for instance, often castigated his opponents as "devotees of ravaging commercialism" intent on desecrating sacred places.[29] Freeman placed himself in a similar tradition, implicitly asking whether any values other than materialism mattered.

Continuing in this vein, Freeman dismissed the copper's economic value. He rehearsed the arguments that favored production—it was valuable; companies were being encouraged to expand production; it would help the war effort—and dismissed them all by pointing out that the war effort and the domestic standard of living would not suffer one bit without an open-pit mine in the Glacier Peak wilderness. Kennecott possessed the right to mine, and the Wilderness Act permitted it, but Freeman hoped the company would recognize that the public interest would not be served by pursuing the mine. (He added that if the mine came to pass, then the Forest Service would require "the highest standards of performance and restoration, under the law.") He confronted the basic paradox of the mine—its legality and the inability of the Forest Service to do anything other than regulate it—and, in doing so, highlighted how his hands were tied.[30]

Since neither spiritual appeals nor the law could force Kennecott to alter its plans, Freeman looked for other means. He noted that the matter was not a case of "either-or" but "rather a case of economics, of choosing alternatives; of balancing a priceless, yet intangible, national treasure against ledger sheets and profits." Making choices—devising compromises—is the art of politics. By laying out the stark choices, Freeman let no one be confused about his position. But since he was relatively powerless, he urged the Sierra Club to "take every possible opportunity to inform the officers and shareholders of the company, and the American public, of the issues at stake on Miner's Ridge." Freeman encouraged action, because he believed that only by taking the issue to the public, who could then take it directly to Kennecott, might the corporation be persuaded to stop its plans. Public activism, rather than government regulation, offered the only way to stop the company that wanted to "gouge out its road" and excavate Plummer Mountain.[31]

Freeman ended in more general, but still soaring, terms. He correctly recognized that Miners Ridge was a test, saying, "In microcosm [it represents] some of the larger conservation issues we face across the continent." The secretary effectively reached his audience, who long had recognized non-materialistic criteria in considering the value of wilderness. Decision-making required long-term perspective and forbearance. "We are a nation bedazzled by technology, and addicted to crash solutions," he said by way of conclusion. "But there are no instant ecologies; no instant wilderness." Our great nation, he feared, stood poised to exercise power without wisdom: "We now have the power, literally, to move mountains. The next few years will determine if we have the wisdom to refrain from doing so."[32] Freeman's words were an eloquent challenge.

Conservationists expressed pleasure and chortled to Kennecott officials.[33] Subsequently, just as Secretary Freemen had avoided mentioning the National Park Service, Kennecott did not counter directly. Instead the mining industry responded in an acerbic editorial in the *Engineering and Mining Journal,* portraying a jousting match between Kennecott and "Sir Orville of Agriculture." Complete with mixed metaphors—"Sir Orville of Agriculture galloped off to war in the Wild West last month aboard a mighty stallion. He was looking for a few dragons to slay."— the editorial informed readers of the secretary's galling call for Kennecott to forgo its Miners Ridge investment. Mocking Freeman for speaking at a wilderness conference in San Francisco—"Pearl-By-The-Bay"—the writer turned serious, recognizing that Freeman's rhetoric could not be dismissed as laughable; its implications were dire. "Sir Orville" meant to close the Wilderness Act's window for prospecting, if not legislatively then through the force of public opinion. The mining industry saw this prospect as a perversion of public interest that was intended to satisfy "a pitifully small vocal minority." The editorial, written in parable complete with an errant knight, surely surprised no readers in its content. Yet its open hostility exposed the thin layer of industry's confidence that covered deeper-seated fears.[34]

All parties recognized that Glacier Peak posed a test.[35] Freeman was clever in meeting the challenge. His presence as secretary of agriculture allowed him to stand in for the federal government and its power of law and regulation. But his department stood powerless before a law that allowed a mining corporation to develop an intrusive open-pit mine in a remote wilderness. By emphasizing this very lack of power, Freeman inspired action by an engaged citizenry on behalf of the public interest. Perhaps even more importantly, he shifted the argument to criteria beyond law and economics to embrace different values, such as beauty and wisdom. Freeman paradoxically used his powerlessness for maximum power. The secretary of agriculture would not be the last person to turn the situation on its head, to use a position of relative weakness to expose the underbelly of Kennecott Copper.

June–July, 1967 23

"An Open Pit, Big Enough To Be Seen From The Moon"

(One possibility, favored by Kennecott Copper Company, for an area of mountains, glaciers, meadows, and forests near here that could otherwise become part of America's greatest national park. Other spectacular local regions are similarly threatened. Only the park can safeguard them.)

This is an "open pit mine" at Bingham, Utah, the largest in the world. The one Kennecott plans for the lovely meadows on Glacier Peak is of the same type, and while not as large as the Bingham hole, would leave a horrid scar which would take more than a thousand years to repair.

AFTER TEN YEARS' preparation, Congressional hearings have begun on an Administration bill that would create a North Cascades National Park here in Washington's "wilderness alps."

As is *always* the case when Congress prepares to set aside an area of such immense beauty for the benefit of all Mankind, there is opposition from commercial interests. To them, a tree is there to be cut, and a great mountain has "value" only for what's inside.

PROFITEERS

Such is the case even here as the day for the creation of this park draws nearer.

A few weeks ago, for example, an ad appeared on these pages (doubtless backed by commercial interests) which claimed a park would "lock out" the people, and would be for "mountaineers" only. It is the same sort of claim the profiteers make about Grand Canyon, Yellowstone, Yosemite and every other place where controls against needless destruction are enforced.

The people who *will* be "locked out" are company people who see trees only as board-feet, rivers as power sources, and the natural landscape as "wasted" if there's no money to be made from it.

Not the only such company is Kennecott Copper, but its case is typical.

THIRTY YEARS OF BLASTING

Within an area officially designated as national Wilderness, and which *should* be part of the park, Kennecott Copper Company is proceeding with plans for an open-pit mine bigger than many of the craters you can see on the Moon.

Mr. C. D. Michaelson, a Vice-President of Kennecott, has said there will "hopefully be thirty years of blasting." As for fixing it up again, he says, "Time will take care of that."

By "time," he means 1,500 to 2,000 years.

But in the meantime? A hole, two thousand feet across and five hundred feet deep, will be gouged out of Glacier Peak Wilderness and 85% of the diggings (the "waste" portion) will be dumped over two hundred acres of terrain.

That is: thirteen thousand two hundred tons of waste, sprinkled over the landscape, every day for thirty years.

(Kennecott says these "tailings" will be kept behind bulkheads so as not to pollute local streams. Where that was done before, waste polluted a lake ten miles away.)

But is this not in a National Forest Wilderness?

DIGGING AND DAMMING

What is not well known to the public is that when the Wilderness Act was first passed by Congress, commercial interests, horrified at the prospect of "losing" all that land that could have been used for digging or damming, managed to have a clause inserted which allowed them in, under certain circumstances.

(Kennecott's excuse in this case is that the national need for copper is critical. In fact, this mine would produce less than a half of one percent of our annual copper consumption, and *not for any present or anticipated need*, but rather to add to stockpiles. Furthermore Kennecott could easily mine elsewhere—it has many other unused mine properties.)

"SECONDARY USE"

As for other equally spectacular areas nearby, whatever the lumber, mining and utility companies want to use, they use. Vacationers, then, get what is left; i.e., "secondary use." They can still boat on the dammed stream, or see the view over stumpland, or hike around the hole where the meadow was.

Only National Parks are completely free of industrial pressures.

CONGRESS CAN ACT

Congress, in acting on the Cascades bill (which presently does not include this Kennecott area) could stop the destruction by including this wilderness within a National Park, together with other nearby regions for which other companies have their plans. (Granite Creek, the Cascade River and Mt. Baker.)

The choices seem clear enough. Is there a great national need for a company's increased profits? Or is there greater national need to protect the few places where a Man can still see the natural world?

"An Open Pit, Big Enough to Be Seen from the Moon." Page from June–July 1967 issue of *Wild Cascades,* newsletter of the North Cascades Conservation Council. Conservationists used this phrasing frequently in referencing open-pit mining. (Courtesy of North Cascades Conservation Council)

Miners Ridge, 1963, with pasqueflower seed heads bursting. Kennecott Copper had not yet publicly announced its plans to develop an open-pit mine just over this ridge. (Courtesy of National Park Service, North Cascades NPS Complex Museum Collection, 16656)

Image Lake with Glacier Peak presiding, 2010. Adjacent to Miners Ridge, Image Lake, is still a popular destination for hikers. (Courtesy of Phil Fenner, photographer)

Plummer Mountain as seen from Miners Ridge, 1963. This photo is from the 1963 North Cascades Study Team investigation. Although Miners Ridge and Glacier Peak were the most common names associated with this campaign, the copper actually lay beneath the surface of Plummer Mountain, seen here beyond the mist. (Courtesy of National Park Service, North Cascades NPS Complex Museum Collection, 11732)

MINING INDUSTRY PLANS TO **RAPE** THE NORTH CASCADES

KENNECOTT COPPER CORPORATION IS THREATENING TO START A $15,000,000 OPEN-PIT COPPER MINE IN THE VERY HEART OF THE PRESENT GLACIER PEAK WILDERNESS AREA AND PROPOSED NATIONAL PARK.

THE REGION'S PRICELESS SCENIC CLIMAX OF IMAGE LAKE AND GLACIER PEAK WOULD BE DESECRATED.

Heated rhetoric from the North Cascades Conservation Council. Anticipating Kennecott Copper's announcement of plans to mine near Glacier Peak, the N3C produced material such as this to fuel the campaign against the company. (*Wild Cascades*, October–November 1966. Courtesy of North Cascades Conservation Council)

"This Is an Open Pit. This Is Not an Open Pit—Not Yet." No regional group was more important than the North Cascades Conservation Council in keeping the Northwest informed of the political developments regarding the Glacier Peak wilderness and mining. This newsletter page was published just as Kennecott announced its plans to mine in the North Cascades and as conservation leaders met with company executives in San Francisco. (*Wild Cascades,* December 1966–January 1967. Courtesy of North Cascades Conservation Council)

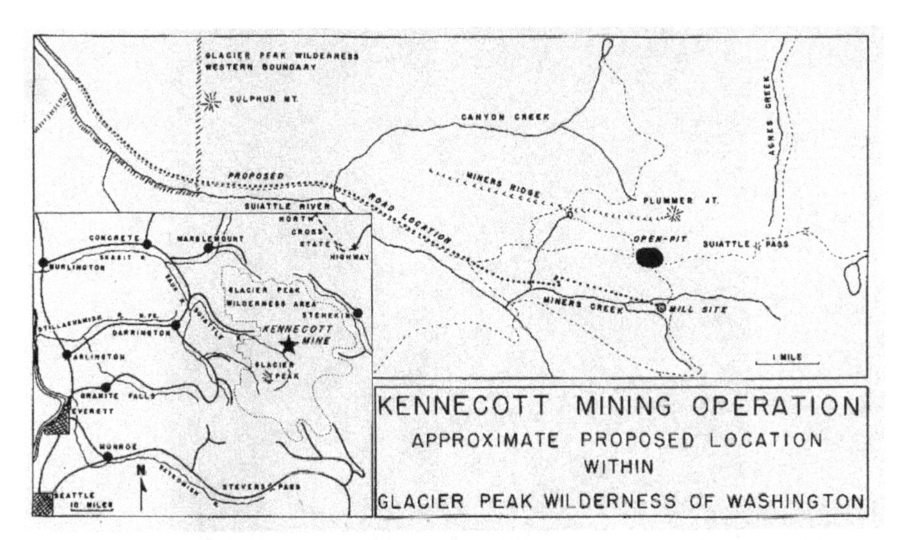

"Kennecott Mining Operation." The North Cascades Conservation Council publicized this Kennecott map showing the corporation's plan, focusing on how the operation would be within wilderness boundaries and among many popular trails. Showing the proposed road to be built, it helped motivate the Northwest outdoors community to act against Kennecott. (Courtesy of North Cascades Conservation Council)

Secretary of Agriculture Orville Freeman (*left*). Freeman was not known as an outdoorsman, but here he is pictured in the Boundary Waters Canoe Area in his home state of Minnesota. The U.S. Forest Service reported to Freeman, and his speech to the Sierra Club's Wilderness Conference in spring 1967 was a rousing defense of wilderness, coupled with a call on Kennecott to adopt restraint. (Courtesy of National Park Service, North Cascades NPS Complex Museum Collection, 11724)

U.S. Supreme Court justice William O. Douglas at 1967 "camp-in" outside Darrington, Washington. Douglas urged protesters to continue their efforts to stymie Kennecott's plans. His presence bolstered wilderness advocates and reflected the national significance of the cause. (Courtesy of Larry Dion / *The Seattle Times*)

President Lyndon B. Johnson signing the bill creating North Cascades National Park, October 2, 1968, a park not including Glacier Peak Wilderness Area or Miners Ridge. *Left to right, immediately behind LBJ:* Lady Bird Johnson (First Lady), Secretary of the Interior Stewart Udall, Chief Justice Earl Warren, Senator Henry M. "Scoop" Jackson (D-WA), Senator Thomas Kuchel (R-CA), Representative John P. Saylor (R-PA), and Senator Frank Moss (D-UT). (Courtesy of National Park Service, North Cascades NPS Complex Museum Collection, 16997)

Mining detritus, 2010, Miners Ridge. In the mountains today, signs of mining exploration remain. These interruptions of natural beauty serve as a reminder of the long-standing threat of mining in the North Cascades. (Courtesy of Phil Fenner, photographer)

Miners Ridge, 2010. This view of Miners Ridge hints at the scale and beauty of this place, dotted with wildflowers and without an open pit. (Courtesy of Phil Fenner, photographer)

SIX | The Doctor

FRED T. DARVILL STRODE into the Biltmore Hotel in New York City holding an oil painting measuring at least two feet by three feet. He surely made an odd impression. The painting depicted one of the Northwest's most iconic scenes: Glacier Peak loomed in the background, Image Lake dominated the foreground, and trees lined part of the lake and reflected on the surface. Darvill had come all the way from Mount Vernon, Washington, a small city sixty miles north of Seattle, where he served as a doctor of internal medicine. Mount Vernon is situated on the Skagit River (now in part a federally protected wild and scenic river), a tributary of which is the Sauk River, a tributary of which is the Suiattle, a tributary of which is Miners Creek, which gathers the rain and snow that falls on Miners Ridge. Just as Darvill saw successive streams connecting his city to the wilderness ridgeline Kennecott hoped to disembowel, that May 1967 morning he planned to link Miners Ridge to midtown Manhattan with his painting. Darvill prepared to disrupt Kennecott Copper Corporation's annual stockholder meeting. His action epitomized an important element of grassroots activism, that of shareholders demanding influence in corporate management. Darvill also represented a growing network of northwestern activists willing to confront the company. At times the defense of Miners Ridge attracted national figures—like the secretary of agriculture (chapter 5) and a Supreme Court justice (chapter 7)—but at its heart were citizens from Washington who found Kennecott's plan unthinkable and its prospect unacceptable. They found ways, creative strategies, to confront corporate power.[1]

Darvill had known a peripatetic childhood as the son of a traveling bookseller but remained well connected to the Pacific Northwest if not fully rooted there. His mother home-schooled him—or rather car-schooled him during sales trips—until high school in Bellingham. He finished high school just as World War II ended and then headed to the University of Washington for undergraduate studies and medical school. He practiced medicine in Mount Vernon for thirty-five years, beginning in 1956. In 1957 he traveled into the Cascade Mountains for the first time, an experience that propelled him into an active outdoor life and sparked a commitment to protect wild places. In 1959, for instance, Darvill rock-climbed at Rainbow Bridge National Monument near Glen Canyon, at that time the site of a central conservation controversy. Construction had begun on Glen Canyon Dam on the Colorado River, which eventually submerged much of Utah's nearby canyon country. Darvill's familiarity with such controversies positioned him to respond immediately and forcefully to Kennecott.[2]

On December 21, 1966, Kennecott publicly announced its plans to mine near Glacier Peak. Two days later the *Skagit Valley Herald* informed readers in Mount Vernon and nearby of the plan, pointing out that the site sat within a potential national park. A few days after that, Darvill wrote a letter to the editor with a succinct message both practical and aesthetic. Current laws failed to protect the North Cascades sufficiently. To counter this weakness, Darvill suggested political pressure in three directions: toward the House of Representatives and Senate to urge the federal government to acquire Kennecott's property; toward the Forest Service to encourage it to "prevent and discourage this nonconforming invasion of a dedicated wilderness"; and toward Kennecott (he provided its address) with personal reactions to the company's plans. Aesthetically Darvill described a singular place: "For those not familiar with the North Cascades, this is the scenic heart land of the entire area, and [it] is without question one of the most beautiful, if not the most beautiful, areas in the entire world." Perhaps he exaggerated, but readers could not mistake Darvill's zeal. At least one northwesterner clipped Darvill's letter and sent it to Senator Jackson, putting the letter into political play.[3]

Darvill also added context and a hopeful note about grassroots organizing. The same year that Darvill had traipsed around Rainbow Bridge, he had floated through the Grand Canyon. Four years later, in 1963, members of Congress, the Bureau of Reclamation, and assorted regional boosters had proposed two dams within Grand Canyon as part of the Pacific Southwest Water Plan to help foster metropolitan growth in Arizona. A sense of vulnerability spread through the West like a flash flood finding every low spot and side canyon. The Sierra Club, with its growing membership, pushed hard nationally to mobilize opposition—through letter-writing campaigns, newspaper ads, congressional testimony, and beautiful publications—gaining enough attention to make politicians rethink their support. Darvill wanted Washingtonians to know these lessons: "Grass roots efforts by individuals were successful last year in preventing the construction of unneeded and unnecessary dams in the Grand Canyon of the Colorado; there is no reason why concerted efforts in regard to preventing extremely defacing mining in dedicated areas in the North Cascades can not be equally successful." He meant to encourage local people and show them that they could stand up to large forces and reverse them. And whether he knew it, Darvill mirrored the Sierra Club: now that the two proposed Grand Canyon dams were no longer a threat, the organization had turned to the North Cascades as one of its two highest priorities.[4]

Beside writing the letter to the editor, Darvill contacted his U.S. representative, Lloyd Meeds, suggesting that the federal government purchase Kennecott's land on Miners Ridge and seeking reassurance that Meeds would defend Miners Ridge. In responding, Meeds offered no reassurance in this immediate aftermath of the Kennecott announcement. Although Meeds agreed with Darvill on one point, saying "we must preserve wilderness values," he did not think an "economy-minded" Congress would appropriate funds to pay a fair price to Kennecott. He noted that copper was selling for thirty-six cents a pound in the United States and fifty-three cents abroad. In other words, copper prices were too high for Congress to buy out the company even if it were interested in doing so. Later Meeds would take stronger stands and seek legislative solutions to preserve the Glacier Peak wilderness, but at this time he told Darvill there was little to be done.[5] So Darvill took audacious action on his own and immediately bought three shares of Kennecott stock for the sole purpose of being able to speak at the company's annual stockholders' meeting.[6]

Fred Darvill served a useful role in the developing campaign against Kennecott: a local citizen not associated with a conservation organization yet willing and able to take a strong public stand against the open-pit mine. In a January 1967 letter to Kennecott he said that as a stockholder he was concerned that if the company proceeded with an open-pit mine in the Cascades, the result might "be a significant deterioration of the corporate image in the eyes of the average American." Such a decline in public perception—a corporate concern conservationists constantly tugged on—would lower the value of Darvill's and other stockholders' investments. He continued, "I would not like to see Kennicott [*sic*] Copper be the first major American Company to blatantly violate the intent and purpose of the wilderness bill, even though such a violation might be legal." The physician recognized the historical import and legal backdrop of what loomed, as Kennecott stood poised to be the first company to test the Wilderness Act's mining exception. As a shareholder from the region around the Glacier Peak wilderness, Darvill put the corporation on notice that local people opposed its plans.[7]

Darvill made further demands. He insisted on both a discussion on the topic and a plebiscite—a formal vote by stockholders—at the company's next shareholders meeting. "I suggest and insist," he said, "that the pros and cons of violating the wilderness system in this manner be submitted to the stockholders for a referendum and vote as their feelings in this matter."[8] Darvill characterized the mine as violating the wilderness *system,* not the *law*—an important distinction that cut through the heart of strict legalities versus the spirit of the Wilderness Act.

Darvill's letter gained the attention of Kennecott president Frank R. Milliken, who responded, telling Darvill that New York law and Securities and Exchange Commission regulations prevented such a vote—something Darvill confirmed with a New York attorney. Although a vote by the stockholders would interfere with management prerogatives and responsibilities, Milliken said he would gladly allow Darvill to make a statement at a shareholders' meeting. He cautioned Darvill that the company was "fully aware" of Darvill's and others' interest "in promoting the cause of wilderness conservation," adding, "We in Kennecott are not unsympathetic to this objective." In closing he stated that even though stockholders did not make management decisions, the company was always "interested in hearing the points of view of our stockholders before reaching such decisions." The fate of Miners Ridge rested beyond Darvill and other stockholders, yet the company's president had responded personally and felt the need to assert the company's sympathy with wilderness, perhaps suggesting that public opinion might shape Kennecott's presentation of its responsibilities. And that was just what Darvill and others who sought to influence American corporations wanted.[9]

Nationwide, currents of discontent and protest were going strong when Darvill arrived at the shareholder meeting in New York in 1967, and the debate over wilderness conservation added one such stream. As a doctor who could afford recreational outings to Rainbow Bridge and a journey to New York while taking time away from his Skagit Valley practice, Darvill represented affluence and liberalism that also characterized the sixties. As others have noted, American liberals after World War II reconfigured their political goals in the prosperous new circumstances prompted by the Cold War. The New Deal in the 1930s had emphasized helping Americans during times of scarcity; liberalism needed an update for times of abundance. Prominent intellectuals like historian Arthur Schlesinger Jr. and economist John Kenneth Galbraith—both of whom advised President John F. Kennedy—criticized American society, focusing on the failure of private wealth to assure the public interest. Too often, they argued, private wealth served only selfish ends while the public interest was ignored. Inequality and poverty were growing in rural areas and cities alike, amid wealth and heavily advertised consumer goods. Liberals sought to counter this misery by promoting investment in schools, roads, hospitals, and parks, not to mention wilderness. The specific strategies of wilderness conservationists grew out of the general liberal movement to balance public quality in a time of private affluence. This included other efforts to challenge corporate dominance.[10]

After decades of deprivation beginning in 1929, Americans could afford to imagine economic restraint, a radical possibility according to capitalist logic. In 1967 the Sierra Club's conservation director, Michael McCloskey, raised this restraint when addressing the Rocky Mountain Mineral Law Institute. "We perceive our relationships in the context of an affluent society," McCloskey explained, "one that produces more than it can consume." This point exemplified staple liberal arguments during the 1950s and 1960s, and McCloskey attempted to explain to the mining industry how conservationists saw the world. Whereas industry believed it served "real human needs," McCloskey maintained that beyond "hard-core poverty," most needs could be satisfied at far lower production levels. Advertising stimulated not simply consumer demand but also crass materialism, as Galbraith charged in *The Affluent Society*. McCloskey encouraged a balancing of "commercial amenities against environmental amenities," prioritizing the latter. He hoped that consumption would decline and result in a more conservative use of nonrenewable resources. But declining consumption meant lower profits for mining and other industries.[11]

Declining consumption might also challenge the postwar political and economic order, perhaps even undermining capitalism itself. Affluence had empowered many American consumers, not only bolstering their bank accounts during much of the 1950s and 1960s but also shaping their sense of political engagement. What historian Lizabeth Cohen characterizes as a "consumers' republic" emerged after the war, premised on mass consumption as the defining economic, political, and cultural feature of the age. From this perspective, government policy and citizen practice fit together hand in glove to promote and preserve the public interest by consuming products. But not everyone saw it this way. The civil rights movement was in full ferment as the consumers' republic ascended, and the interaction of the two gave rise to a consumer protection movement, with protests aimed at protecting the public from shoddy, unsafe products and rectifying discriminatory, predatory practices.[12]

Grassroots organizing characterized these movements, and Saul Alinsky became one of the most prominent organizers who focused on marginalized peoples. Active since the 1930s when he had started working in Chicago's working-class Back-of-the-Yards neighborhood, Alinsky developed a set of organizing tactics widely adopted by such poor communities.[13] His 1945 *Reveille for Radicals* furnished a blueprint for "the building of people's organizations," including by-laws for such a group.[14] For the next generation, the self-professed radical published *Rules for Radicals: A Practical Primer for Realistic Radicals*, an updated 1971 handbook for tactical organizing against power. In it Alinsky highlighted the 1967 effort with FIGHT (Freedom, Integration, God, Honor—Today), an organization that challenged Kodak over its discriminatory hiring policies toward African Americans. FIGHT organizers recognized that an economic boycott of Kodak would likely fail, given the company's market dominance and the expected loyalty of the local community in Rochester, New York, the company's headquarters location. So they deployed a new tactic of invoking the power of shareholder proxy votes to change corporate priorities.[15] For example, Alinsky would accept speaking engagements and get shareholding organizations to assign their votes to FIGHT as a proxy. They especially targeted churches to leverage the symbolism. Once FIGHT wrangled votes in this way, it accrued power, and its activists could more effectively make demands at shareholder meetings, gathering media attention. This happened in late April 1967 when a group from FIGHT disrupted a Kodak meeting. Although the board of directors did not acquiesce to FIGHT's demands that day, a settlement occurred shortly thereafter. FIGHT and Alinsky used the proxy tactic effectively to increase national attention to a problem and to tarnish the corporation's public image. The tactic served its purpose and inspired others, like Darvill, whose status as the Washington State vice president of the Federation of Western Outdoor Clubs hardly suggested radicalism.[16]

An older form of corporate protest to demand greater accountability also inspired actions like Darvill's, epitomized by brothers Lewis D. and John Gilbert, who interrupted corporate proceedings effectively for more than half a century. Beginning in 1933, when he first rose to ask a question at a shareholder meeting and was ignored, and continuing until 1992, Lewis crashed corporate meetings. "Gilbert Goes to Meetings and Raises Roof," a *Newsweek* headline from 1937 put it.[17] But he was not really crashing, because he owned small numbers of shares in more than fifteen hundred corporations, allowing him to legitimately voice his concerns and call for greater corporate accountability, which he termed corporate democracy. This became the Gilberts' strategy: openly advocating for the rights of corporate owners—the stockholders. By some accounts, Lewis transformed corporate stockholder meetings from occasions where fewer than a dozen sat in silence to events with a thousand or more articulating demands or asking probing questions. Over the years Gilbert provoked members of boards to lose their tempers; some came after him "fists revolving" or invited him into an alley afterward. For forty years the Gilberts published an annual account of stockholder activity at corporate meetings. The *New York Times* called Lewis Gilbert the "most prominent—and indefatigable—of the professional corporate hecklers," a role he could play as an independently wealthy man whose Park Avenue residence put him two blocks from Central Park amid America's most powerful people and institutions. The American Research Council once characterized Lewis as "deeply conservative, his only predilection being a militant belief that since shareholders own the corporations in which they have invested they should have a voice in the management of their own property."[18] This perspective drew on a widening base.

In fact, the Gilberts' strategy worked mainly because of increasing American affluence, as measured by rising numbers of stockholders. The *New York Times* reported in 1967 that the number of Americans who held corporate shares had more than tripled in the previous fifteen years from 6.5 million to 20 million, reflecting growing prosperity despite inequality and poverty. This meant three times as many corporate owners to potentially weigh in regarding the management of their companies, like the Gilberts. Energized by the growing consumer movement, many shareholders protested. The same *Times* article described several interruptions at shareholder meetings. One corporation had faced a shareholder who wanted to know what percentage of the company's profit could be linked to the Vietnam War. Another had demanded to know whether any corporate contributions had gone to CIA front organizations. Anticommunists at another company had protested because the company bought raw materials from communist Yugoslavia. The greatest disruption came at Eastman Kodak's meeting when African American shareholders "interrupted proceedings with angry demands for more jobs." Even though "[these examples of] discontent were too scattered to have much influence on management," the *Times* reporter characterized the potential of such actions as "immense." Still, shareholder protests amounted to a "heretical idea" because these "dissident owners" maintained that values other than profit ought to govern corporate policies.[19]

These tactics of shareholder proxy voting and protest highlighted corporate vulnerability in the public view. As long as annual shareholder meetings remained hidden and stockholders stayed home, management remained basically private. But as soon as groups organized or arrived to speak on these occasions or vote on corporate policies, then what had been no-news events became newsworthy. In some ways, then, the Gilberts' and Alinsky's activities may have resembled stunts. Writing of Alinsky, the *New York Times* reporter hinted as much: "Certainly, he can make next year's annual ritual livelier than ever." But then he added that Alinsky had the potential to affect corporate governance: "It is unlikely that he can overthrow profitable management but with the backing of affluent members of the civil rights movement, particularly churches, pension funds or foundations that are big shareholders, he should be able to make them sit up and take notice."[20] Such was the power of proxy campaigns and challenges to corporate autonomy. They tapped into a broad range of supporters who used their economic power as leverage to move corporate priorities.

These variants of the consumer movement offer a window to Darvill's approach, even though he was not going after a company for racist hiring or seeking legal redress because of unsafe consumer products. His action drew the same breath as the consumer movement. The record does not show whether Darvill knew of the Gilberts, but he enacted their values perfectly. Darvill advised the Kennecott board about his stock property, but he also spoke about the public property wilderness represented. Most protests in the 1960s, especially in popular memory, occurred in public—at southern lunch counters, in university quadrangles, on the streets—but corporate boardrooms also were sites of protest. This historical moment needed and used all types and sites of protest, and they were evident in the Glacier Peak campaign. They arose paradoxically in part from the growing affluence that American consumerism had brought. What soon came to be called environmentalism grew out of this milieu, concerned about quality of life more than mere material standards of living.[21]

Darvill prepared for his moment. He worked with local and national activists to generate publicity and coordinate plans, suggesting that organizations arrange for "a photographer from the New York Times and other publications friendly to the conservation cause" be present for the shareholder meeting. He believed that if the media published and broadcast images of what he planned to bring—a painting or a photograph—this might generate "public pressure opposing the open pit mine." Once the doctor even proposed to the Sierra Club's Mike McCloskey that they encourage a "set of pickets" on the sidewalk outside the meeting hotel.[22] On Darvill's behalf, Brock Evans, writing then for the Federation of Western Outdoor Clubs, contacted *Medical World News* to recommend that they cover Darvill's action and background on the campaign: "Public awareness of the problem is essential if this desecration of natural beauty is to be prevented."[23] Darvill recognized that he needed to get the attention of the Kennecott board but also that attention to the issue could be magnified if national media covered the story.[24] He would not be the only one speaking out at the meeting either. Sierra Club president George Marshall appointed his proxy to another Sierra Club member to share objections.[25] Kennecott knew opponents were coming but not what they would say or how it all would play out.

"I have come here today to talk about wilderness and beauty," began Darvill at the appointed moment—May 2, 1967, at ten in the morning in the Bowman Room of New York's famed Biltmore Hotel. Darvill continued and spoke eloquently—an "impassioned talk," according to one report—about Glacier Peak wilderness. With his oil painting and a photograph in hand—symbolically bringing the Cascades' wild mountain peaks to Manhattan's towering skyscrapers—the doctor opened with what surely was an uncharacteristic comment for Kennecott Copper Corporation stockholder meetings. He strove to make Glacier Peak vivid for the board, to make it a real place and not an anonymous source of ore and profit. "To reach this area at the present time," he said,

> one must first follow a trail through a deep valley of virgin forest for some eight miles; then one climbs five miles on switchbacks upward for a vertical mile to the ridge crest. Here, in the alpine meadows, there is an unexcelled panorama of Glacier Peak towering 10,500 feet above the forested floor of the Suiattle Valley. . . . This ridge extends about five miles; the whole ridge is an alpine meadow carpeted with a profusion of wild flowers intersperced [*sic*] with the arrow-like silhouette of the alpine fir.
>
> This ridge is the scenic climax of [the] entire North Cascades, America's wilderness alps. They are perhaps the most beautiful mountains in the world, and most observers agree that the area of Miner's Ridge and Image Lake is the acme or heart of the area.

In referencing "scenic climax" and "America's wilderness alps," Darvill relied on the script conservationists had developed over years of activism for Glacier Peak and the North Cascades. He continued, explaining how hikers came, "drawn by the unexcelled scenery, to experience wilderness and to saturate themselves with beauty." "To walk this ridge," Darvill said, "is an experience to be recalled for a lifetime. . . . Dozens of different species of wild flowers grow waist-high along the trail. Small streams plunge downward into the valley below." Such descriptions of glorious mountain scenes gave concrete examples of the nature of Miners Ridge to the board, to other shareholders, and ultimately to the public. Darvill's words transported his audience from the Biltmore to the Cascades and, he hoped, from a boardroom mentality to a trailside reverence.[26]

After describing Eden, Darvill moved to Original Sin. He listed the intrusions Kennecott's production plan entailed: a 350-acre open-pit mine, a concentrating plant, fourteen miles of new road. Then Darvill named some who publicly opposed the mine: Secretary of Agriculture Orville Freeman, Governor Dan Evans, Representative Lloyd Meeds, and Seattle newspapers. As he summarized it, "high officials in the government and the man on the street" wanted nothing to do with Kennecott's mine. In tallying this diverse cast, Darvill hinted at various objections and obstacles. The cabinet secretary promised that the nation's war effort would not suffer without the mine. The governor explained that his committee studying the North Cascades saw no reason for it and pledged to use "all the forces available" from the state to oppose it. The member of Congress indicated that a bill would be forthcoming to buy Kennecott's claims to "prevent the defacing of this pristine landscape." And a bill in the Washington legislature proposed to tax open-pit mines. Darvill also indicated that the Snohomish County Board of Adjustment's zoning regulations prohibited the mine. He warned of the pressure that would follow—"if necessary at the level of the courts"—if the company asked for a zoning variance. This litany reminded the corporation how, while its mining engineers probed for ore, conservationists mobilized popular and political opinion and explored legal avenues. Just before concluding his remarks, Darvill seemed to remember that he was a stockholder and there to do more than inform the board of opposition. He said he hoped that Kennecott would not become known in popular opinion as "a desecrator of the wilderness" by creating an open pit as "a monument to the rapacity of man." Finally, the doctor ended with a bit of doggerel, perhaps designed to be easily quoted in subsequent press reports: "Let it not be said, and said to your shame, that all was beauty here, until Kennecott Copper came." And, with that, Darvill completed his stockholder duty.[27]

Darvill was, of course, not the lone voice that day. Kennecott described the meeting as a "lengthy discussion," while the *Wall Street Journal* reported that the discussion of the "controversial proposal" took up "much of the meeting."[28] Two others spoke in favor of wilderness, including Abigail Avery, who owned a cabin in the Stehekin Valley, and George Marshall's proxy, who read a letter from the Sierra Club president.[29] Marshall noted that the nation "would survive very well without this mine, but [it] would be immeasurably poorer if it [lost] this scenic area," and then asked for a five-year moratorium to give time for the public to buy Kennecott's claims. Kennecott president Frank R. Milliken rebutted, after reporting "record levels" of copper production and sharply rising profits, mainly asserting that the Wilderness Act allowed the planned mine. Milliken also claimed that all the nation's wilderness areas combined amounted to an area twice the size a New Jersey, a slight exaggeration meant to suggest that one small mine could not damage the entire wilderness system. A less diplomatic response came from Everett Conlon, a New York stockholder: "Beauty is in the eye of the beholder. They say God put the mountains there for us to enjoy. Well, maybe he put them there for us to get the copper out." This brash remark showed that Darvill's view of sacred wilderness did not sway everyone. Another opponent, a Mrs. Daniel Weinstein, offered a less provocative defense of the mine, simply calling for "the greatest good for the greatest number," long a conservationist slogan often hauled out to justify economic development. (Incidentally, one of the Gilbert brothers also attended the meeting after gathering nearly 1.5 million proxy votes, just shy of 6 percent, in an unsuccessful effort to vote for a limit to annual pension payments for the corporation's executives.)[30]

In the meeting's immediate aftermath, Darvill questioned whether his action had mattered. He told the *Seattle Times*'s Walt Woodward, a stalwart mine opponent, that he hoped the board had gotten the message. However, he doubted whether it made a difference. "I am left with the impression that this corporation strongly wants to go ahead with the mine," he said, though he added, "maybe what was done today will shake it up a bit." According to Woodward—but appearing nowhere else in the record—Darvill suggested that the company "make a gift of the area to the people as the 'Kennecott Copper High Cascades Park,'" indicating that the doctor not only was committed but also possessed a rich sense of humor.[31]

The board did not indicate whether it was shaken up or had gotten the message. However, a small opening appeared from the company, and Darvill's gambit rippled beyond the Biltmore Hotel. President Milliken said that Kennecott still awaited final work on various "engineering and economic studies, now underway."[32] This equivocation contrasted with a report just a month earlier stating Kennecott would begin work on the mine "as soon as the snow melts."[33] But more important was that the story had legs.

The *Washington Post* ran a story on the day of Darvill's stockholder meeting speech and an editorial the following day. Neither referenced Darvill, but the two seem coordinated. The editorial, "Crisis at Miners Ridge," conceded the possibility that someday the nation might face a dire copper shortage that justified "leaving a ghastly scar in the place of natural beauty." But nothing justified it in 1967.

The *Post* moved from the specific landscape of Glacier Peak to general principles and expressed shock that corporate calculations, not public agencies, determined whether to mine on public lands—something Secretary Freeman's recent speech, as well as Darvill's statement, had publicized. The newspaper explained the ways that the mining exception in the Wilderness Act allowed corporate interests to outweigh public interest. Deciding whether a crisis merited destroying wilderness should be made, the *Post* said, "by a responsible national agency rather than a private company interested primarily in profits." Policy questions on public land logically required a public agency to decide, the *Post* maintained. To rectify this problem, there should be two amendments to the Wilderness Act. First, the secretary of agriculture ought to be authorized to condemn mining claims within wilderness. Second, if that failed, the highest priority would be enforcement of the Wilderness Act stipulation that mining be done "only under conditions compatible with preservation of the wilderness." Besides offering specific policy solutions, the *Post* clearly favored what it described as "one of the most alluring samples of the unspoiled manless world." In other words, one of the nation's leading newspapers, printed not far from the Capitol, preferred wilderness. Also, it brought national attention to Miners Ridge but rightly explained that this particular problem represented a systemic shortcoming in wilderness legislation. Seen this way, Darvill's actions, which seemingly prompted this editorial attention, helped bring national attention for stronger wilderness protections. Miners Ridge again tested the nation's understanding and tolerance of the Wilderness Act with a little prompting by a Mount Vernon doctor.[34]

Fred Darvill pushed forward the campaign against Kennecott working from within the corporation, albeit as a holder of only three shares, to challenge management priorities. Those three shares, owned for not even six months, were enough to gain Darvill a platform at the annual meeting. And that platform attracted national media attention to what was at stake at Miners Ridge and in the nascent National Wilderness Preservation System. When Darvill brought a painting of Image Lake and Glacier Peak into the Biltmore Hotel, he helped knit together the nation's landscapes—the wild Northwestern mountains and the corporate boardrooms of New York City—with the emerging protest politics. The effect of Darvill's efforts added to those of the preceding month's Sierra Club address by Agriculture Secretary Orville Freeman clarified and amplified the debate. That a small-town doctor from the Skagit Valley could successfully bring national attention to the debate suggested the resonance of the issue. And, after all, Miners Ridge was public land, the nation's land, a place where private decisions needed to be weighed against public priorities—at least according to Darvill, Freeman, and the *Washington Post.*

Next, a Supreme Court justice would take to the trail, where he would meet with student protesters and longtime Northwest activists.

SEVEN | The Justice

IN A FORESTED RIVER valley where the road ended at a Forest Service campground, two hundred or so people, maybe more, gathered near midday in mid-summer 1967. More than two dozen miles from the nearest town, the campground hosted what newspapers dubbed a "camp-in" or, more expansively, a "camp-out, hike-in, love-in, be-in."[1] Given the counterculture rising then, such an event might have suggested a place to drop out for a while, to escape the turmoil of politics in troubled times, to leave worldly things behind. But these people had come there precisely to engage. Joining the young and old who converged under the trees was William O. Douglas, a justice on the U.S. Supreme Court, marking this camp-in as something different, something residents of nearby Darrington eyed with muted hostility.

That Douglas was there was not surprising—he had done such things before—but it signaled the rising importance of publicity to the fight against Kennecott and the ability of this cause to attract highly placed and networked political figures. But Douglas brought substance, not just publicity and connections; he did not just fly in for a speech. He knew Miners Ridge and had spent decades already working for wilderness. Thus, he brought with him some moral authority derived from firsthand experience of place and political processes. Further, he represented legal authority and lent the cause credibility, even legitimacy.[2] Douglas also reinforced and extended the critique activists already deployed against Kennecott. As he had been before, the justice was a public intellectual for conservation.[3] Serving as a public intellectual, though, sometimes meant being a lightning rod for controversy and attracting protesters. Because of his presence, Douglas illuminated divisions between local people and communities on the one hand and public lands and national interest on the other. The woods on the edge of Glacier Peak Wilderness Area resembled a rustic town hall meeting with a circuit judge presiding.

Born in 1898 and raised in Yakima, Washington, Douglas stood as the most prominent northwesterner in national life and arguably the most visible public figure endorsing conservationist agendas.[4] Douglas had risen from modest means with a deep intellect and unstoppable ambition, gaining academic success and then appointments in the Roosevelt administration's Securities and Exchange Commission. Roosevelt named him to the Supreme Court in 1939 when he was only forty, making him the second-youngest Supreme Court justice in history. The court—a full-time job for most—did not contain Douglas's restless mind, so he traveled the world and commented on political and legal affairs in books that reached wide audiences.[5]

Although Douglas spent his career in the political world of Washington, D.C., he possessed an abiding interest in the natural world and turned to environmental causes. Throughout the 1940s he built a reputation as something of an outdoorsman, drawing on his familiarity with the Cascades from his many youthful tramps and climbs through the range.[6] In the 1950s he deployed that identity and energy toward two well-publicized protest hikes, one just outside the nation's capital and the other on the Olympic Peninsula in Washington State. In both cases Douglas protested National Park Service plans to build roads along a scenic corridor—a defunct canal on the East Coast and a remote beach in the Northwest—to lure auto tourists. And in each case Douglas drew compatriots, including conservation activists and journalists, while he hiked several days and many miles to generate enthusiasm for hiking over driving, trails over roads, slow over fast. The attention the justice brought to the cause helped tip the scales toward the wild.[7]

Most of Douglas's wilderness time came in intimate parties of a few people, after which he shared insights and inspiration in popular books and articles. Such was the case with Glacier Peak. In the late 1950s Douglas hiked to Miners Ridge with a childhood friend and a local packer. Such trips had become commonplace for Douglas during his summer time off the bench, after which he wrote up accounts that shared the natural history, personal stories, and political challenges each place inspired.[8] The places he visited comprised a list of endangered wildernesses that Douglas used to create narratives for teaching ecological lessons, understanding management threats, and inspiring political change.[9]

This role helped the wilderness movement in the 1950s and early 1960s as it found its footing. Douglas brought public attention to it through his actions and his writing. When Douglas published his book chapter on Glacier Peak in 1960 in *My Wilderness: The Pacific West,* Rachel Carson had not yet galvanized a popular movement for ecology, Congress had not yet passed the Wilderness Act or the National Environmental Policy Act, and Earth Day was still a decade away. Consequently, as a prominent public figure, Douglas introduced and interpreted American conservation in ways accessible to audiences unfamiliar with the locales he visited and described and the problems he identified. Glacier Peak fit this pattern perfectly.

For Douglas, in addition to their ecological value wild places served critical cultural functions. As the nation became increasingly "apartment-born"—a phrase he used frequently to connote overpopulation, urbanization, mechanization, and laziness—citizens needed "testing grounds" that nature provided to ensure "strength and character." Douglas and others like him worried about how Americans depended too much on machines. Without the "elemental challenges that sea and mountains present," he wrote, they would not "experience the sense of mastery of adversity." Douglas's ideas reflected a commonly expressed belief about the need to be challenged by the natural world, a belief often connected to desire for masculine mastery over environments. But he also valued the simple escape, "so remote from cities and factories and farms," that wilderness pockets allowed. For him, rising populations meant that only the mountains and sea could furnish the serene respite needed to reset. Douglas went so far as to call it "physical and spiritual therapy" and compared wilderness areas for humans to what wildlife refuges were for ducks or antelope.[10] Glacier Peak, not even seventy-five air miles from downtown Seattle, offered an ideal wilderness break from modern hassles.

Yet Douglas rehearsed more than these familiar and general arguments, because he wrote as the Forest Service determined Glacier Peak's wilderness status. Douglas and his companions worried. On the hike he describes in *My Wilderness: The Pacific West*, they rested at a point where they saw two vulnerable spots, both unprotected by the Forest Service: Miners Ridge, where "[m]ining interests want to chew . . . up in search of copper," and below it, where Douglas anticipated "[t]he gutting of Suiattle Valley" by loggers and road-builders. The threats to this place were heightened because, as Douglas diagnosed it, the "area is so little known it has few friends." Friends and firsthand experience were essential to conservation; people do not destroy what they love, in Douglas's view, and knowledge of Glacier Peak would create love.[11] Douglas functioned effectively in the role of publicist, helping raise broader awareness.

The justice's solution to Glacier Peak was simple: protect it and its surroundings as a wild area. Douglas appealed to readers, who he hoped would press their representatives and the Forest Service:

> If the valleys and ridges of this Glacier Peak area are sealed from commercial projects, we will have forever in America high country of enchantment. Those who search them out will learn that an emptiness in life comes with the destruction of wilderness; that a fullness of life follows when one comes on intimate terms with woods and peaks and meadows. Then a person learns that he was born not to destroy the chain of life of which he is a part but to nourish it; that he owes respect not only to his elders but to the snow buttercup at his feet and the hoary-mantled marmot, whistling to him from a rockslide.[12]

This was the power of wilderness. With a legal designation, wilderness barricades commercial threats. As a natural place, it could reframe one's perspective to demonstrate the interconnectedness and equality of all life. Douglas showed the practical and the spiritual aspects of wilderness. He would do the same at the 1967 camp-in.

Douglas published *My Wilderness: The Pacific West* in 1960, on the cusp of important political change. He spoke for a growing number of Americans dedicated to the outdoors, some of whom had already proved willing to transform environmental questions into public campaigns, but the full-fledged environmental era complete with new federal statutes and a mass movement had not yet taken flight. The Wilderness Act was still four years away from passage. Rachel Carson's clarion call in *Silent Spring,* which later inspired a grassroots ecology movement, was still two years from publication. *My Wilderness,* then, came at a moment when it could be celebratory, and it mostly was celebratory, even though Douglas inserted pointed political barbs. The conflict between wilderness advocates and their opponents that we take for granted today had not yet hardened. But opposition that greeted Douglas in the North Cascades in 1967 exposed fault lines that subsequently fractured further.

It was not the first time Douglas had been greeted by opposition on the trail. In 1958 the justice had joined other wilderness luminaries to protest a road proposed on the Ocean Strip of Olympic National Park. The Wilderness Society had just held its annual meeting at Stehekin, on Lake Chelan, in the heart of North Cascades country, and while so many conservationists were in the state they staged a protest hike on the coast. Polly Dyer, so instrumental in the Glacier Peak wilderness campaign, organized the hike that drew Douglas, Howard Zahniser, Wilderness Society cofounder Harvey Broome, and National Parks Association president Sigurd Olson, among others. The beach hike represented a key moment in the emerging postwar Northwest environmental movement, and part of what made it endure came at its conclusion.[13]

After three days on the coast, covering more than twenty miles, Douglas clambered off the beach to meet two protesters: a man and his young son, each holding two signs. The man, Larry V. Venable of Port Angeles, directed the Automobile Club of Washington and staked a position for automobile tourism. The placards held by his son—too young to drive and in photographs barely visible behind the large signs—clearly articulated the importance of roads: "We Own This Park, Too. We Want a Shore Line Road" and "Fifty Million U.S. Auto Owners and Their Families Like Scenery, Too!" For Venable and others who favored the road, it promised to increase tourism by providing easier access. How to integrate automobiles in parks' recreation purview was a question that had long plagued the National Park Service generally and Olympic National Park particularly.[14]

But the Venables carried another message to protest Douglas and his crew's presence. A third sign read "Bird Watcher Go Home," belittling the seriousness of conservationists' efforts. "Flower picker" was another epithet used to disparage an aesthetic appreciation for nature, and, in coded language, questioned men's masculinity.[15] "Bird Watcher Go Home" suggested that the only serious engagement with nature was an economic and mechanized one and that watching birds or hiking along beaches was elitist or effeminate or both. Despite wanting the road for tourism (and surely some tourists would marvel at eagles as they rode in their Chevrolets), Venable felt safe there in logging country hinting at the superiority of working, rather than walking, in the woods. His objections were intended to weaken the conservationists' case like a river eroding a streambank and to subvert any moral authority they possessed.

Venable's sign also worked to undermine the protesters' legitimacy by questioning who should be allowed to speak on behalf of nature—not bird watchers, certainly, and not "outsiders." The dynamic of local residents versus outsiders was long-standing, dating back to when government agencies first began regulating the previously open range. Enclosing local common spaces disrupted existing economic patterns, and when done by a federal agency it frequently generated resentment.[16] To Venable, Douglas and the rest of the hikers amounted to outsiders who had no claim on that landscape. "Go home," his sign demanded. But conservationists by that time had developed and strongly articulated a vision of public lands as belonging to all the people in the nation.[17] Douglas frequently wrote about such places as the inheritance of all the people.[18] So to the wilderness advocates, they *were* at home already—at home on public lands, in the wild, open to all.

Almost a decade later, in 1967 for the camp-in, Douglas traveled to the North Cascades at the invitation of David Birkner, who led the Statewide Committee to Stop Kennecott Copper. The opposition that met him there would be somewhat greater than the father-son team on the coast, and the opposition's concerns would be more developed.[19] In the lead-up to the camp-in, rumors circulated about potential violence, an overblown notion that provided dramatic tension for journalists more than it dampened the event. Throughout its existence the town of Darrington had supported mining and logging in the Cascades, resource-dependent activities apt to be economically unstable. Although the Summit Timber Mill still operated in Darrington, other mills had closed recently.[20] Thus the news of a possible open-pit mine cheered residents of this town tucked along the Sauk River and surrounded by mountainous national forests. Hopeful residents greeted the camp-in protesters with a sign: "Welcome Kennecott."[21] Drawing on the emerging script from similar towns, local citizens saw the protesters as "outsiders [who] want . . . a private playground" and said that working communities would become "dead towns" without infusing main street with new industry. Local millworkers welcomed the idea of more workers and money in their town and spoke protectively of it as "our home country."[22] A Seattle newspaper quoted Chuck Neidigh, a steely-eyed Darrington logger, who castigated conservationists as just a "bunch of birdwatchers . . . trying to take something away from those who have to work for a living." The rhetoric was virtually the same as that Douglas had faced at the end of his 1958 beach hike.[23] But Neidigh extended a more pointed critique than Venable by suggesting that conservationists disrespected those who had to "work for a living."

Historian Richard White had this sort of confrontation in mind when he wrote his now-classic essay, "'Are You an Environmentalist, or Do You Work for a Living?': Work and Nature."[24] White argued that environmentalists failed to reckon seriously with how people worked in nature, too often condemning those whose physical labor most closely shaped nature—like a logger in Darrington or a miner who might come to live there. Instead, environmentalists—including those gathering up the road from Darrington—tended to focus on nature as a site for leisure while their own labors tended to be distant from the mountains and rivers that provided the natural resources on which they depended. Seeing the ways everyone's work tied to nature might lead to more interesting and responsible engagement with the world. The refrain about people who need to "work for a living" has followed wilderness enthusiasts in the Northwest for half a century and more, marking a fault line between those who know nature through their work and those who know it through play. The logger Chuck Neidigh announced clearly his perception of disrespect from conservationists fighting for wilderness, and, at least in part, he was asserting the value of the work he did on the land.[25]

Workers were only one constituency that confronted the mine protesters. State representative Henry Backstrom also crashed the party, arguing that Kennecott's mine would "be a shot in the arm for the working people." The company would leave the area with a road and a "man made lake after they're through," cheered Backstrom, a Democrat from nearby Arlington, who assured listeners the he wanted to "preserve the beauty of this area as much as the next man."[26] The politician suggested that Kennecott would improve nature, echoing arguments made commonly by engineers and others that it is humans' role to finish or fix creation.[27] As a policymaker, Backstrom perhaps had the long view in mind. However, anticipated jobs and development would be short-lived. Construction of the mining operation would last only three or four years, and mines do not support renewable resources or long-term jobs.[28] As for the "man made lake," a filled-in pit could not compete with Image Lake reflecting a majestic Glacier Peak during a summer dawn.

Ultimately, these counter-protesters would be few and impotent. The Sierra Club's northwest representative, Brock Evans, characterized the conflict as "grossly overplayed," with Seattle newspapers, more than anyone else, playing up the confrontation.[29] One observer, Robert Michael Pyle, a young University of Washington student active in the Conservation Education and Action Council that helped organize the protest, recollected the counter-protest as "a bit cliché" and not at all nasty.[30] So the circumstances of environmental protest and counterprotest on the Pacific Coast and in the North Cascades mirrored each across the years. However, the extractive nature of Kennecott's proposed mine entailed economic identities in ways the proposed park road along the beach had not. In both instances Douglas listened to protesters but ignored their root concerns to speak on a different plane. Instead of jobs, the justice aimed at materialism and morality.[31]

And so it was that on August 5, 1967, "a band of 150 adults, kids, dogs, and an assortment of people wearing beards and beads," as the press dubbed it, converged on Sulphur Creek Campground, about thirty miles east of Darrington. The group of all ages milled about, signed petitions, and ate lunch amid the Douglas fir forests where Sulphur Creek spilled into and mingled with the milky, glacier-fed waters of the Suiattle River. After a while, the protesters headed up the trail a short distance to listen to Justice Douglas speak. Pyle later recalled that he "felt in the presence of greatness."[32] Relying on themes familiar to those steeped in American history and nature writing, Douglas said, "You've seen the wilderness and its glories and its wonders and its beauty and you may be the last generation to see it . . . the frontier is just about gone." In the next breath, drawing on his narrative from *My Wilderness,* he railed against the materialism symbolized by a corporate mine: "Let's not turn everything into dollars. Let's save something for the spiritual side of man." He continued to plead: "From the city fathers down we must start thinking and planning to keep this country beautiful—to keep the mountains wild and undefiled."[33] Photographs published in newspapers soon afterward show an impassioned Douglas, eyes and open hands raised to the sky, beseeching the audience huddled around him holding signs protesting the mine and its foreseen impact on the Skagit River system. According to Pyle, Justice Douglas "made everything seem possible."[34] Douglas's presence there in the North Cascades not only brought needed attention to the conservationists' cause but also boosted activists' morale and hopefulness. He also made clear to Kennecott that regional resolve remained strong.[35]

As Douglas spoke there in the "sunny dappled shadow along the Suiattle," reporters captured his words, which he later expanded on.[36] As a symbol of ultimate legal authority, Supreme Court justice Douglas acknowledged the legality of Kennecott's proposed action. The Wilderness Act plainly allowed prospecting and mining, but, as he told his audience, "just because something's legal doesn't necessarily mean it's right."[37] Douglas's comments reflected his long-standing interests in justice and conservation, retooled specifically for this campaign.[38] Instead of urging a lawsuit or targeting legal loopholes—actions increasingly common among activists as Congress passed environmental legislation in the 1960s and 1970s—Douglas pushed the protesters to "try to appeal to the collective conscience of the corporate community represented by Kenn[e]cott Copper, to change the minds of stockholders, directors and others who are interested in the corporation, so that the corporation will not become identified in the public eye with the vandalism of the kind that will result if this open-pit operation goes into effect." Similar to Fred Darvill's appeal to Kennecott stockholders, Douglas aimed at the corporate conscience, using moral suasion. Events like the camp-in might prod that conscience.[39]

In his remarks Douglas alternated pessimism with optimism. He rooted his pessimism in corporate callousness and materialistic values so dominant in American society. Too often in society, he lamented, owners directed "their eye only on the dollar sign in disregard of all the spiritual values of the outdoors," a practice that meant that America was "destined to become a very barren, ugly bit of wasteland." The places there under threat—Glacier Peak, Image Lake, Miners Ridge—constituted "a part of our great wilderness heritage," he said. "If we lose this battle with . . . Kenn[e]cott," he feared, "we stand to lose many other battles." The choice was clear to him and his audience, and success depended on two things—and here is where his optimism showed. First, the nation's constitutional system enshrined majority rule but also protected minority interests, something the justice had labored for three decades to secure. In this case, the nation had to protect the "rights of minorities . . . who like to backpack, ride horseback, to climb, to escape the noise and din of civilization and find the solitude and peace that is available only in the remote wilderness areas." Second, an aroused public might "appeal to the community's sense of justice, the whole corporate community, the whole body politic" and declare its preference for beauty over "a few paltry dollars." So hope lay in politics—both the government and grassroots citizen action—as well as a moral sense of right. That combination, in Douglas's view, proved cause for resounding optimism.[40] The campaign against Kennecott depended on it.

Justice Douglas's fealty to the U.S. system of government and faith in democratic action made him an ideal symbol for wilderness activists. He inspired and reminded them of their power to inform their government and shape its policies, while linking the cause to fundamental American themes. As the 1960s unrolled, the Forest Service and Park Service faced each other in their final North Cascades showdown. Central to that were citizens, who stood up in hearing rooms and wrote letters for the record to advocate for wilderness or mines, parks or multiple use. Douglas made his reputation—in conservation and civil rights circles—with a strong belief in constitutional rights and the voice of Americans, as his presence and words reminded those gathered on the trail in 1967. Douglas had long advocated for the public to be heard in conservation issues.[41] He reflected the values of the grassroots activists and helped prime them to influence the legislators who came to the region to understand what northwesterners wanted in a North Cascades National Park.

National figures like William O. Douglas and Orville Freeman could generate headlines, and atypical actions like Fred Darvill's could rally attention, but when the time finally came to resolve the stalemate at Miners Ridge, citizen voices mattered, as Douglas knew.

EIGHT | The Park

IN MID-SEPTEMBER 1968, THE U.S. House of Representatives passed its version of Senate Bill 1321 with small changes. The Senate agreed to the House version, and on October 2, President Lyndon B. Johnson signed into existence North Cascades National Park, located north of Miners Ridge. This achievement culminated years of patience and persistence on the part of many activists in the Northwest and beyond. In the nation's capital, no one had worked harder than Scoop Jackson, the chair of the Senate Interior Committee. When Congress agreed to the bill, *Seattle Times* columnist Walt Woodward sent a congratulatory note via Jackson's secretary: "My very best to Scoop and hooray for a good guy who really accomplished something. Now . . . let's figger out a way to keep Kennecott from gouging up the landscape, huh?"[1]

The latter remark, in Woodward's colloquial voice, signified several relationships in the just-completed park campaign. Woodward recognized that despite the great accomplishment in protecting millions of acres of North Cascades landscape, some places remained vulnerable, at risk of "gouging." Now was not the time to rest. The momentum built over the previous years—a decade and more for conservationists and half a dozen for national politicians—needed to be maintained to finish the job. Woodward also understood that North Cascades National Park and Miners Ridge were connected. These places had become intertwined in the popular mind. Although ultimately the park did not include Miners Ridge, the controversy over Kennecott Copper Corporation's proposed actions galvanized the public to support the park. Understanding Miners Ridge or North Cascades National Park requires understanding both.

Strong public opinion had pressured the Forest Service, out of whose lands the national park had to be pried. Fred Darvill, William O. Douglas, and Orville Freeman had in preceding months publicized a particular threat—an especially egregious one—to North Cascades wilderness. Their words and actions had raised awareness of Glacier Peak wilderness, the threats to it, and the legal shortcomings for protecting one of the nation's premier wild places. This knowledge brought together regular citizens to take up the cause that committed activists built. Jackson explained the necessary process clearly, telling park advocates how the system worked: "I can't *give* you a park. . . . But if you get up a big enough parade, I'll step out front and lead it on in."[2] And there is plenty of evidence that Jackson cared about this landscape, writing once, "The North Cascades area has a special personal importance to me. Some of the most memorable experiences of my boyhood were gained tramping its trails. I find it hard to escape the feeling of proprietary interest in its fate."[3] The proprietary feeling was familiar to many Northwest conservationists besides Jackson, who recalled summer weeks among the North Cascade peaks. Several sets of hearings in the Northwest gave the public an opportunity to voice their perspectives, furnishing an important historical record that shows antipathy to Kennecott, increased distrust of the Forest Service despite its efforts to bolster its wilderness credentials, and growing forcefulness in the nation's—especially the northwestern—conservation movement. The metaphorical trail to North Cascades National Park had gone through Miners Ridge and been blazed by citizens whose joint efforts had influenced legislators and agency officials alike. To be sure, the campaign's leaders drew from white professional classes, but their growing numbers of supporters confronted powerful corporate resistance in Kennecott and government recalcitrance in the Forest Service and among some legislators. Citizen voices against Kennecott's open-pit mine at Miners Ridge amplified the national park crusade. After Miners Ridge was not included in the park, as the columnist Woodward reminded Senator Jackson, the work needed to continue.

Almost since the founding of the National Park Service, citizen groups had called for it to take control of the North Cascades, irritating the Forest Service. As the 1950s moved into the 1960s without the agencies resolving these competing claims, conservationists moved in and raised the stakes. Many northwestern conservationists clung to hopes that the Forest Service might create a suitably protected wilderness area for Glacier Peak, but the Sierra Club began pushing for a national park. In 1958 David Simons, a student at the University of California, Berkeley, talked the Sierra Club's David Brower into hiring him to survey the North Cascades. Simons worked with gusto, producing powerful reports that proposed a 1.3-million-acre Lake Chelan–Glacier Peak National Park. Several organizations began working to support such a park, including the National Parks Association, North Cascades Conservation Council, and Sierra Club. This was a transitional period, with fluidity among various actors and organizations and their different proposals as they worked hard to protect wildlife and wild places. The Forest Service increasingly appeared as a target, and its lack of responsiveness pushed many, including Brower, to finally seek salvation in the National Park Service.[4] In the *New York Times* in 1959, Brower called for a national park for the North Cascades.[5] Such advocacy fanned the flames of bureaucratic rivalry, as *Times* columnist John B. Oakes noted. According to Oakes, "The Department of Agriculture is often involved in contradiction and conflict in conservation," blocking NPS efforts to protect "irreplaceable scenic or wilderness areas." Among the many such areas over which the agencies fought, none was more important than the North Cascades. Oakes pointed out that the Forest Service's efforts up to that point had been anemic compared with what the region's scenic beauty deserved.[6] In 1960 Thomas Pelly, a Republican member of Congress from Washington, introduced House Resolution 9360 charging the Department of the Interior to investigate a national park for the North Cascades.[7] In 1963 Mike McCloskey compiled "Prospectus for a North Cascades National Park," the length of a short book, which presented a compelling and detailed case for a park.[8] Such advocacy presaged the national push.

Bowing to popular and political pressure, the so-called Treaty of the Potomac, an attempted working agreement between the Forest Service and the Park Service offered a solution. Announcing a "new era of cooperation" concerning outdoor recreation, Secretary of Agriculture Orville Freeman and Interior Secretary Stewart Udall informed President John F. Kennedy that, among other things, the departments would put aside their rivalry and conduct a joint study with members from both departments who would resolve "in an objective manner" the jurisdictional conflicts in the North Cascades. The study would determine how to manage and administer federal lands in the North Cascades in a way that would "best serve the public interest."[9] The North Cascades Study Team consisted of two members chosen by the secretary of the interior (Owen S. Stratton and George B. Hartzog Jr.), two chosen by the secretary of agriculture (George A. Selke and Arthur W. Greeley), and a chair (Edward C. Crafts) selected jointly. These experts visited the North Cascades and held public hearings in the Northwest, which provided an early snapshot of public sentiment concerning its wild landscapes.

Management of public lands required public input, generating a long record of public opinion. Just before the study team began its work, conservationists expressed frustration about shortcomings of the hearings process. Sierra Club president George Marshall wrote to Northwest conservation stalwart Harvey Manning:

> I agree with you that it is most unfair, and essentially undemocratic too, to have Forest Service hearings frequently at out-of-the-way places for most who would like to testify—and on working days at that. This procedure is also symbolic of an essentially wrong approach to wilderness—that the opinion of those living close to it is more important tha[n] that of the 180 million Americans to whom the wilderness areas belong and who have a right to enjoy it as wilderness if they so wish.[10]

Wilderness, not to mention national parks and forests more generally, constituted public space, a national landscape kept open for the American public whether they lived on the forest boundary or in New York City.[11] Local communities obviously enjoyed a particular relationship to nearby public lands, one that often expressed itself as proprietary, but their sense of entitlement was not rooted in law. The perspective Marshall shared cut to the heart of critical paradoxes that governed public land issues.

As the North Cascades Study Team circulated in the region to investigate the landscape of mountains and politics, they stopped to listen to the public in two Washington communities—Mount Vernon (home of Fred Darvill) and Wenatchee—in October 1963. The hearings found the public there much in favor of maintaining the Forest Service's multiple-use approach, fearing adverse effects to the regional economy without it and loss of places for hunting. Then the team listened in Seattle to an audience the team characterized as "conservationists, professional people and individual citizens," who clamored for the park or more wilderness. Altogether, of the 216 who testified in all three hearings, 126 favored continuing USFS management, with a minority of 87 favoring more wilderness, a national park, or changing multiple-use priorities. However, written comments offered a counterpoint, with a ratio of roughly four to one in favor of creating a national park. Only 408 of the 2,375 written testimonials wished the Forest Service to continue its management practices. These data were suggestive. Because the strongest support for wilderness had long come from cities, it was not surprising that hearings in the smaller communities nearest the national forests yielded support for the USFS, while Seattle favored more recreation. This difference in those who wrote to the study team reveals something of the growing organizational power wilderness advocates marshaled and the region's changing economic structure. Local groups like the North Cascades Conservation Council effectively mobilized its members, and their perspectives informed the study team as it wrote its recommendations.[12]

When the group finalized and submitted *The North Cascades Study Report* in 1965 it recommended significant changes—twenty-one in total—to the region's public lands administration. Although the changes reached from Mount Rainier to the Canadian border, the recommendations that mattered most to this study centered on the existing Glacier Peak Wilderness Area and the North Cascades National Park then newly proposed. The team advised adjusting the wilderness area boundary in several places, including the Suiattle River corridor, but though this change inched toward Miners Ridge it did nothing to affect mining one way or another. The park proposal also incorporated nothing related to mining near Image Lake; that absence still mattered and set a stage for future debates. In short, the study team recommended numerous administrative changes but none that affected the status quo for the minerals embedded in Miners Ridge.[13]

The study team report, dated October 1965, actually appeared in January 1966, being made public at a large Seattle press conference with all the principals and more present.[14] The next month, Jackson's Senate Committee on Interior and Insular Affairs held hearings over two days and gathered written testimony from hundreds. The resulting record included more than a thousand pages that ranged over a wide array of positions and concerns. Despite the study team report's only general references to mining, industry representatives used the subsequent public hearings as an opportunity to stake claims on critical access to crucial mineral deposits both specific to Miners Ridge and in general, citing the tradition of multiple-use resource management.

As did timber companies, mining interests viewed public lands as a storehouse of resources to be developed economically for public benefit via private profit, and they spoke at the public hearings to reiterate this position and ensure that their privileged position continued undiminished. To mining corporations, the Forest Service's multiple-use approach made sense, especially when paired with the utilitarian ethos that permeated the agency's history. Mining companies identified this practice as part of the "national tradition" of economic development of public lands. Any move away from this "greatest good for the greatest number" philosophy in the form of wilderness or park designations violated their sense of progress, and so they urged delay on such new legislation until a "more enlightened definition . . . regarding priorities" appeared—a common tactic of delay for more study seen throughout the history of wilderness designation.[15]

In addition to defending mining amid traditional management regimes, mining groups justified their activities as temporary, small, and negligible. Mines created small footprints, they claimed, hardly noticeable. But from those small geographic areas came great value. In short, minerals were the "least visible" resources but ended up becoming "the most valuable of all," as a representative from the West Coast Mineral Association put it. And even in those small areas, mining interests tried to persuade their audience, the damage of mines was overstated and did not deface scenery as opponents charged or would be "concealed or gradually healed" with modern techniques. So these proponents rejected changes to the status quo and found criticism of mines overblown.[16]

Finally, mining advocates identified important reasons for promoting mines in the North Cascades, specifically at Miners Ridge, of which national security was foremost. The Northwest Mining Association representative who attended the hearings had worked for the U.S. Bureau of Mines during World War II, when he had urged developing U.S. copper mining, and thought now that the Vietnam War demanded this long-delayed project.[17] In 1966 many in the nation feared a copper shortage, and the Department of Commerce placed controls on exports to keep the valuable metal available. Clearly every possibility to increase domestic production needed to be pursued.[18] Another witness, a geologist who had worked at Miners Ridge for more than a decade with Kennecott's subsidiary

Bear Creek Mining Company, reported a 30 percent increase in "free world demand" over the previous decade. "Our national security is dependent upon an adequate domestic mineral supply," he asserted. By framing the issue in terms of national security, he communicated a message that Senator Jackson, as a strong cold warrior, would have understood and appreciated.[19] The mining industry spoke with a clear voice, united by a desire to ensure that legislative and administrative priorities would preserve its rights to mine. Testimony from extractive industry representatives showed opposition to parks and wilderness in sometimes practical, sometimes principled, and sometimes self-interested ways. Increasingly, though, their voices fell out of tune with the public.[20]

The study team hearings occurred almost a full year before Kennecott announced its plans for Miners Ridge, so the mining corporations and their conservationist opponents jousted less over the place and more over general principles. The locale was rarely mentioned in the public debate at this time, although some arguments specific to the Glacier Peak wilderness began that would continue. Many people testifying at the Senate hearing about the study team report urged the National Park Service to take over Glacier Peak Wilderness Area.[21] In the context of the study team, prior to Kennecott's announcement, these ideas came more from a position of celebrating Glacier Peak's scenic qualities than protecting it from Kennecott's drills and dynamite. Another idea floated at the time was Brock Evans's call to appropriate federal money to "purchase the copper mining patents in the very heart of it, on Miner's Ridge."[22] Many would adopt and advocate for this idea in the coming years, but in early 1966 Miners Ridge was folded in to the larger and more general struggles over the North Cascades landscape, a struggle that soon focused on specific park details.

If the study team launched a ship from its harbor, it was far out at sea by early 1967. President Lyndon Johnson announced his support for a North Cascades National Park in a special message to Congress on January 30, 1967, called "Protecting Our Natural Heritage."[23] By March 1967, Washington's senators, Scoop Jackson and Warren Magnuson, had introduced S. 1321 after Interior Secretary Stewart Udall delivered a bill approved by both Agriculture and Interior Departments.[24] Having a bill under consideration laser-focused Northwest conservationists. The legislation proposed a new national park, new national recreation areas, and new wilderness areas, but it left Miners Ridge in the status quo. Nevertheless, Kennecott's open-pit threat to Miners Ridge, now formally announced, galvanized many to the causes of wilderness and the park.

After introducing the bill, Jackson convened public hearings again—a practice that contributed to democratic governance, assuring that the public would at least get to speak out in an attempt to shape policy, and an unsurpassable source in furnishing a historical record. Northwesterners wrote to Jackson to ask to speak at the hearings. In those short notes, they expressed genuine dismay at

Kennecott's threat. A student activist from the University of Washington, for example, asked for time for someone from his organization, the Conservation Education and Action Council, to speak at Northwest hearings. The activist, Robert Michael Pyle, went on to become a noted environmental writer and even then could turn a phrase: "We urge you to do all in your power to halt Kennecott; the snows are melting and the disaster is impending."[25] Protecting wilderness *always* meant racing to forestall the forces of extraction. Kennecott's plan to begin mining as soon as the ground cleared of snow heightened the sense of urgency evident in Pyle's plea.

These hearings on the proposed park bill offered northwesterners their first opportunity to air their perspectives in public after Kennecott's consequential announcement. Consequently more people were engaged with the question of North Cascades wilderness and, in particular, how Miners Ridge and Kennecott fit in that larger regional conservation puzzle. Whereas only a handful of people addressed the disruptive potential of Kennecott's copper claims in the 1966 study team hearings, the next year saw that number increase dramatically. In the Senate's 1967 hearings before the Subcommittee of Parks and Recreation more than 330 people spoke or wrote directly about Kennecott or Miners Ridge, and 95 percent opposed mining in the wilderness. Roughly one-third recommended including Miners Ridge in a new North Cascades National Park and about one-quarter mentioned buying out Kennecott's claims. Sometimes concerned citizens suggested both. Nearly one in five indicated that they had been to Miners Ridge or nearby, lending an air of personal experience to the proceedings and simultaneously showing how the place inspired those who had never visited. While Kennecott developed its plans, conservationists mobilized, and the testimony advanced their argument.[26]

Those speaking at the hearings recognized the high stakes involved at Miners Ridge and understood that Kennecott's planned mining represented a test of the Wilderness Act's strength and thus transcended Glacier Peak. In this way, Miners Ridge was a specific place, where many people felt as though Kennecott was attacking not only the land but also them personally. It also represented something general concerning the larger wilderness movement. Speaking at length for the Sierra Club, Michael McCloskey clearly articulated this broader import of Miners Ridge. McCloskey ticked off a number of lingering concerns with S. 1321, drawn from the deep knowledge he had developed when creating the "Prospectus for a North Cascades National Park." But, like many others, for greatest impact he ended with Miners Ridge, a climax of sorts that highlighted the most egregious shortcoming from the draft legislation. An Oregonian with serious Northwest credentials, McCloskey reminded the senators in attendance that conservationists had long called for Miners Ridge to be in a national park.[27] If it were to be in the national park, the obvious incompatibility of open-pit mining and wilderness grandeur would generate sufficient popular outrage to stop the action. Like the Forest Service, the National Park Service could not rescind preexisting property or mineral rights, but it had stronger administrative capabilities to prevent mining than did the Forest Service.[28] If the new park did not protect Miners Ridge, as the initial version of S. 1321 indicated it would not, McCloskey hoped "public officials [would] be no less vigilant in using every available authority to protect publicly owned lands and resources within the Glacier Peak Wilderness." Despite enjoying important protections against timber-cutting and road-building, wilderness still needed help to prevent desecration by open pit, so McCloskey pressed policymakers and the federal land agencies: *"The adequacy of our laws, the wisdom of our administrators are . . . on trial at Miners Ridge. So also is a sense of responsibility to the public of a private corporation."* These two sentences express the principal issues more succinctly than any other source. McCloskey did not mention the Wilderness Act or the Forest Service by name, but those clearly were in his sights. His framing focused on three points: the weakness of existing wilderness laws, the discretion and wisdom of administering public lands, and the relationship between public and private responsibility. Each was complex and problematic, and together they formed an almost insoluble puzzle. McCloskey understood better than most the legal, administrative, and political aspects of wilderness—and how they acted together. McCloskey's presentation to the Senate subcommittee reached the peaks but also recognized the underlying bedrock.[29]

Another speaking out at the hearing, Polly Dyer, also conveyed much from what she had learned over the preceding decade. No one had been more central to Northwest wilderness activism.[30] Dyer reminded the audience of the history of the Glacier Peak wilderness. During World War II many had been eager to gain the copper from Miners Ridge, but somehow the Allies defeated the Axis powers without the copper. Surely, Dyer implored, the same would hold true again. Her point about restraint resonated with wilderness activists, because the Wilderness Act at its core was about restraint. Existing land laws—"archaic," to her mind—too often privileged profit. Dyer urged Congress to "reconsider its giveaway of the public domain . . . when extraordinary scenic and intangible values are at stake." As she saw it, there exist "some lands in these United States that should never have become the private holdings of any one person or company—and this is one of them; there are some lands that should be held in trust for all Americans—and this is one of them." Dyer spoke like many conservationists by elevating some lands over others. Besides vaunting this place and castigating outdated laws, Dyer dismissed those who cited dire needs for copper. "If copper were in critical supply for a last ditch effort in the Nation's survival, copper would not now be used in new plumbing, cooking utensils, or quarters," Dyer declared. Until then, she implied, leave Miners Ridge alone. For Dyer, who certainly spoke for many others, the situation in the Glacier Peak wilderness was intolerably rooted in outmoded ways of valuing and managing land.[31]

Dyer and McCloskey agreed on that and were among the most notable figures in a crowd that saw this not only a test of the Wilderness Act but also as a referendum on the Forest Service's sputtering wilderness management. Many who wrote or spoke to the senators found the agency's approach insufficient and urged resolution. The Forest Service's inability to protect wilderness convinced many to advocate relocating the administrative control of Miners Ridge to the National Park Service. These partisans characterized the Forest Service as too compromised, especially by its history of supporting logging and minimizing wilderness values. The multiple-use policy so touted by Forest Service personnel (and mining and logging interests) was seen as a sham by those who found no way to reconcile open-pit mining and wilderness in the same locale. The agency, they were convinced, possessed neither the will nor the power to stop Kennecott or similar threats. Some urged the Forest Service to test its power, to regulate Kennecott as stringently as possible until perhaps a court decision overruled. But most conservationists, knowing the agency's history, did not trust the Forest Service to do even that.[32]

Part of the Forest Service's weakness came from the Wilderness Act, not the agency itself. Kennecott spotlighted those gaps. "Obsolete mining laws" unjustifiably privileged mining corporations in too many places, according to one witness who offered a common refrain. If it could not stop Kennecott, then the Wilderness Act required revision. Citizen conservationists, state legislators, and Christian ministers, among others called for legislative reforms, but in the meantime Glacier Peak and Miners Ridge remained vulnerable, which worried them. Abigail Avery, who lived in Massachusetts but also owned land in the Stehekin Valley, spoke for many: "I feel that the fact that . . . Kennecott can consider this [mining] is a glaring weakness of the Wilderness Act, and I won't want the Glacier Peak Wilderness Area to demonstrate and dramatize the weakness of the wilderness bill." Avery was prepared for compromise. She did not rush to judge the Forest Service; she held out hope that the agency might rein in the mining company. Yet if the agency could not prevent another disaster like the one at Holden, Washington, then she would throw her support to the National Park Service.[33]

The frequent reference to Holden during the Senate hearings showed how conservationists positioned themselves in both geographical and historical context. Abandoned by Howe Sound Mining Company in 1957, the Holden mine lay only about ten miles to the east of Miners Ridge, over the Cascade crest, close enough for many to know both sites. In the decade after the mine closed, pollution from it lingered, especially in Railroad Creek, worrying recreationists. Railroad Creek flowed out of Lyman Lake, visited by many who also visited Image Lake as part of one trek.[34] Anglers and hikers reported no fish in the creek. For opponents of Kennecott, then, Holden offered a prime example of failed multiple-use policies by the Forest Service. It symbolized the havoc left behind after mining. One Bellevue man put it simply, "Having seen the disgusting mess on Railroad Creek at Holden left by the Howe Sound Mines I would hate to see the same crime committed in the Glacier Peak Wilderness." Northwesterners who loved the Cascades saw what could go wrong, a grim reminder of the accumulative effects mining left on the environment. They were not exaggerating; the site today is still undergoing Superfund-mandated remediation.[35]

Besides placing the planned Miners Ridge mine in local context, many understood it as a national issue and part of a wider conversation. Importantly, national and local concerns coalesced. Local organizations informed national allies, a committee of correspondence in action, as William O. Douglas described it once.[36] And experiences of mining, of Kennecott, of conservation victories flowed back to the Northwest. Many, for instance, referenced the famous Bingham Pit operated by Kennecott outside of Salt Lake City as a monumental example of open-pit mining, a "canker sore" in the words of one woman from a Seattle suburb.[37] Beyond comparing open-pit mines, some who voiced concerns at the 1967 Senate hearings mentioned articles—and, more importantly, photographs—that appeared in publications like *National Geographic.* Writers like Paul Brooks and articles in the *New York Times* were referenced too. All this evidence showed how Miners Ridge occupied a part of a *national* conversation and demonstrated that national publications and writers and activists incorporated the Northwest into their beat. This scale was something that Northwest conservationists used to their advantage; doing so was key to challenging the opposition.[38]

The printed record of the hearings, 1,100 pages, conveys the voices of many northwesterners who were saddened and enraged by Kennecott's audacious plans to invade a wilderness and irrevocably alter it. From teenagers to retirees to soldiers in Vietnam, from those who knew Miners Ridge firsthand and those who never would, this broad constituency shared attitudes and ideas about beauty and restraint, the importance of public input and long-term planning, and more.[39] But the most common theme, perhaps, was the notion of *incommensurability.* The term is ponderous, certainly, but its meaning—the inability to be compared—matches this situation. Letters and statements kept returning to cost versus benefit, although not as an economist employed by Kennecott might have used such analysis. For lovers of the mountains, this was more than a simple calculation of costs and profits. More needed to be accounted for in this figuring. As one Seattleite asked Kennecott's president in a letter included in the hearings record: "Why don't you realize that the relatively small amount of copper gained here would be immeasurably overshadowed by the terrible damage caused? Damage not only to the scenery, but to the name of your company?" Again and again came the refrain: this was too much sacrifice for too little copper.[40] In the hearings, the campaign boiled down to this essential idea of incommensurate sacrifice. Certainly, many criticized Forest Service management or legal inadequacies of the Wilderness Act. And others denounced Kennecott's short-term, profit-driven calculus as leading to nothing short of raping the land—a common, if troubling, characterization.[41] Of course, many shared their love of the place, extolling the area's special beauty and recounting life-changing experiences in the wilderness. But in the end, for so many who testified, the issue exposed the mismatch between the irrevocable loss of wilderness and what they saw as a paltry gain in the amount of copper to be mined. What would be lost simply could not compare with what might be gained.

After the hearings concluded, Senator Jackson's mailbox kept filling, and most correspondents shared their general objections to extractive industry in wild, scenic places. More nuanced approaches did exist, though, including that of William P. Jeske, a Seattle man who built on the idea of incommensurability in a way that highlighted the economic mismatch from another angle. Jeske liked the Senate's park bill, because he thought it balanced preservation and multiple use; however, Glacier Peak's absence from the park and lack of protection for it alarmed him. Having hiked through the area the previous summer, Jeske judged it "easily the most beautiful [area] in North America, possibly anywhere in the world." The mine threatened to destroy it and impoverish everyone. Quick to assure Jackson that he was not one of those "ultra-conservationists who object to cutting a tree or moving a stone to build a highway," Jeske explained that copper mining constituted a different sort of problem, an incommensurate violation. Copper mining was not part of the regional economy and thus did not support a large population base. But what most griped Jeske was that mining companies exhausted resources and left behind messes, just as Howe Sound Mining Company did at Holden. Corporations worked not for the public interest but for their own self-interest, Jeske wrote, "and I have no quarrel with this—it is as it should be under the free enterprize [*sic*] system." Yet these companies paid less than full costs when mining on public lands, because they left tailings and pollution behind. "Why should I pay this cost?" asked Jeske. "If the copper is so valuable that it can pay its full cost, then I wouldn't object to the mine. But I do object to a public subsidy." Jeske exposed the off-balance economics of public land extraction: private profit and socialized costs. Even those who did not mind multiple use found Kennecott's plan flawed.[42]

The Senate hearings exposed public sentiment, which overwhelmingly favored parks over pits, protection over profit. Jackson recognized the upswelling of opinion regarding Miners Ridge in particular. His Senate colleague, Republican George Murphy from California, forwarded an emphatic plea from a Los Angeles constituent who urged legislative action to stop Kennecott from its "devastation." Jackson told Murphy how the hearings had contained "considerable testimony" about Kennecott and Miners Ridge, saying that it was "important that enlightened public opinion be brought to bear on this issue in support of efforts to protect the public interest in this magnificent area."[43]

Although Jackson was considering additional legislation to enlarge the park to include Miners Ridge, the campaign required still more traction to gain sufficient support. Thus, the third stage in the park campaign began after the work of the study team and its hearings and after the Senate bill and its hearings. Now the North Cascades Conservation Council, Sierra Club, and regional allies worked to convey their ideas and views to legislators to sharpen the park proposal and show their political strength. If it had not been a popular movement before, it became one as 1967 moved into 1968 and as the Senate yielded to the House.

Attention gathered and the pace quickened to create the North Cascades National Park. The Sierra Club and the North Cascades Conservation Council had built a strong foundation during the previous decade and came to this moment with a sense of history. In a sharp-looking pamphlet, the Sierra Club informed the public that ten great national parks had been proposed between 1890 and 1950, with nine subsequently becoming parks. The tenth—the Ice Peaks National Park, proposed in 1937—had never reached congressional debate.[44] Although a mere 20 percent of the proposed Ice Peaks Park, the North Cascades proposal before Congress promised to conclude this unfinished business. The club found fault in the Senate bill and offered corrections. Prominent among them was corralling Kennecott by transferring the Glacier Peak Wilderness Area to the National Park Service. This plan followed N3C's 1963 proposal and "would provide much-increased protection for this scenically superlative terrain." Broadly, the Sierra Club wanted further additions to wilderness areas in the region, but for Miners Ridge they knew that only national park status would prevent the open-pit mine since wilderness law had failed.[45]

The pamphlet spread throughout the region and indeed the nation. And so did the Sierra Club's full-page ad said to have appeared in the *New York Times* and *Los Angeles Times* with its heading "An Open Pit Big Enough to Be Seen from the Moon" and its photo of Bingham Pit.[46] The imagery followed a strategy deployed earlier by the North Cascades Conservation Council, which had put two photos on the front of its *Wild Cascades* newsletter. On top was a terraced, open-pit mine, looking like a desolate moonscape, with inset words: THIS IS AN OPEN PIT. Beneath that, a photo depicted rocky peaks in the distance, partly covered with snow, with a meadow and trees in the foreground, with inset words: THIS IS NOT AN OPEN PIT—NOT YET.[47] The national newspaper ad extended this visual argument. Text accompanying the horrifying image described all sorts of terrible Kennecott plans: thirty years of blasting, tens of thousands of tons of waste, and unquantifiable pollution. Then, point by point, the Sierra Club countered every reason Kennecott offered for the mine.[48] At the bottom of the page, coupons of a sort were pre-addressed to the House and Senate Interior Committee chairs and Washington's senators and governor, with a blank one for a U.S. representative, making it easy to clip and mail them. Their message: "I believe the North Cascades deserve better protection. A national park would provide the best protection possible. For this reason I support the administration's bill for a North Cascades National Park but urge that it be expanded to include other nearby areas needing protection. Please take whatever action is necessary to protect the Glacier Peak Wilderness from being despoiled by an open pit mine." These coupons were dutifully clipped, signed, and mailed, piling up in the offices of decision-makers like Scoop Jackson as evidence of the growing public sentiment.[49]

These campaigns worked, building momentum for the park throughout and beyond the region. And Kennecott had fostered this movement. The Sierra Club's Northwest representative, Brock Evans, jokingly wondered to a correspondent whether "we have friends in Kennecott who thought up this proposal to get national attention to this area."[50] This is unlikely, of course, but Evans made the useful point that the threat to Miners Ridge—outside the park boundaries—bolstered enthusiasm for strong preservation efforts in the North Cascades. Kennecott provided N3C, the Sierra Club, and others with prime motivation for extending the vision of North Cascades conservation.

The efforts of conservation leaders, acting like potent fertilizer, strengthened the grassroots. Not only did northwestern activists and hikers understand the issues, they spoke out about them. N3C and the Sierra Club made it easy for them to send letters, and the study team and Senate hearings gave them opportunities to testify. By the time the House Interior Committee met in Seattle in April 1968 to hold its own public hearings, regional activists buzzed. Evans recalled the Seattle hearings as "the real crystallization of the environmental movement in the state of Washington."[51] The moment, long in coming, felt momentous; locals believed the park lay within reach.

But the committee chair, Wayne Aspinall, represented a formidable obstacle. The western Colorado politician was a deliberate legislator, so careful and methodical that one journalist dubbed him "molasses-moving Aspinall." This predilection collided with conservationists' sense of urgency, and activists' environmental agenda was not necessarily among Aspinall's highest priorities. To move on the Wilderness Act, Aspinall demanded that Congress create the Public Land Law Review Commission, which he then chaired from 1964 to 1970, to assess the proper balance between resource development and preservation. Aspinall enjoyed a well-earned reputation for favoring development over preservation, such as supporting the pending Colorado River Basin Project that would re-plumb the interior West. In the spring of 1968 Congress faced not only the North Cascades park bill but also the final stages in the Wild and Scenic Rivers Act, the National Trails System Act, the Redwood National Park Act, and more. All of the negotiations had to go through Aspinall, who sat like a dam with the only lever to open the spillways. The pressure of this environmental flood irritated Aspinall, who felt unable to control the legislative agenda. Compounding his problems, Aspinall had a prickly relationship with Scoop Jackson. More than anything, Aspinall sought water security for the Southwest and gazed with covetous eyes toward the Northwest's abundant rivers, imagining a massive diversion to satiate the dry Southwest. Jackson rebuffed such schemes. More generally, Jackson had a more expansive legislative vision than Aspinall, who thought senators with their six-year terms did not have to be in touch with their constituents closely. So Aspinall arrived in Seattle distracted and pressured, with perhaps less generosity in reserve than normal.[52]

And a firestorm greeted Aspinall. The hearings were scheduled for the Benjamin Franklin Hotel in a room with a 130-person capacity, but 800 people showed up and overwhelmed the committee. The scene irritated Aspinall. The controversy had been "blown all out of proportion," he said. He blamed Jackson and the press for whipping up the public into an agitated state—the press because it "promotes controversy" and Jackson because he had run an "irresponsible" Senate hearing in 1967. In Seattle, Aspinall fumed: "I don't know who these people are. Are they hippies or part of a Seattle drive to get out into the country?" Surely some hippies and other student activists attended. But so did doctors and lawyers and engineers and housewives. To accommodate the crowd, Aspinall split the committee and held simultaneous hearings in two rooms. Even that proved insufficient, so attendees drew lots for the opportunity to testify.[53]

Anyone paying attention to the crescendo in Northwest conservation circles would not have been surprised by the testimony. The tune stayed the same, but in spring 1968 it was played louder than ever. Dozens and dozens mentioned Kennecott, and virtually all of them opposed the corporation with the same arguments expressed at the Senate hearings. The Seattle setting might have meant that only an urban sensibility would be represented, but people came from other areas of Washington to speak. One of them was William A. Rivord from the small timber town of Sedro-Woolley, a place destined to be a gateway to the park. Rivord had spent his career amid "foresters, fish and game men, and loggers." He hunted and climbed and hiked through the North Cascades. "For a long time I have wondered how much of our great heritage we will leave for future generations to use and enjoy," he told Aspinall. "My hope rested in the passage of the Wilderness Act until Kennecott Copper Co. proposed to dig a mammoth hole near Image Lake on Miners Ridge. This proved that classified wilderness areas under Forest Service management are vulnerable to exploitation and disfiguration. Miners Ridge is littered with many 5-gallon gasoline cans which are rusting away. Steps should be taken to prevent further mutilation of this magnificent area." Rivord's concerns resonated with many and represented the views of countless others who were appalled that rusting cans and mammoth holes were legal in wilderness. Aspinall wanted to take responsibility and correct the record, though, telling Rivord not to "blame the Forest Service area because of the mining enterprise that you suggested because the Forest Service area cannot do anything about it. That happens to be part of the provisions of the Wilderness Act." He added, "Blame Congress. We are the ones to blame. Do not blame them because their administration is in accordance with the provisions of the act." Perhaps testimony like Rivord's made a difference, though. The spectacle of the hearings seemed to influence Aspinall, who assured the public afterward that he would "do his darndest" to get the bill out of his committee before the end of the congressional session.[54]

In the aftermath of the House hearings, the stakes were clarified. Part of the crowd at the hearings included opponents of the park who had mobilized for what they rightly saw as a last chance to stop it. Still, park and wilderness advocates outnumbered them three to one, according to Patrick Goldsworthy.[55] One of Seattle's dailies characterized Aspinall as having had a "snit" in the city. The paper noted that Aspinall's "intemperate comments" only heightened the controversy that he complained about. From the perspective of *Seattle Post-Intelligencer* editors, the North Cascades were "of extreme importance to the citizens of this state and the press would be failing its responsibility to give it less than extensive coverage." It was time for the House to follow the Senate and pass the bill.[56] Even papers outside the region noticed the Seattle hearings, with the *Washington Post* declaring it an "event of national significance."[57]

And so it was. Back in D.C., Aspinall and Jackson parried, threatened, politicked. In his N3C history, Harvey Manning deployed his characteristically colorful words: "The two high lords of the American earth, secure in their citadels, one commanding the Senate, the other the House, exchanged thinly veiled threats."[58] Such was the system, and the system worked. By fall, the House and Senate had passed the bill, and on October 2 a beleaguered President Johnson, then in the waning days of a troubled presidency, signed the bill, saying that the law preserved "for the pleasure of these people one of the beautiful regions on God's earth."[59] The act was flawed, from conservationists' perspective. They objected to too many timbered valleys remaining under Forest Service control and thus subject to logging. But they had achieved something important, if symbolic: they had wrested nearly seven hundred thousand acres from the USFS, which showed the mounting power of grassroots activism. Brock Evans liked to say he would have settled for a park the size of a table just to show the Forest Service that political power existed to challenge its own.[60] Moreover, the park primarily was wilderness, with adjacent national recreation areas as buffers, a new concept in NPS management.[61] The Sierra Club and David Brower received a lot of credit for the park as well as for the simultaneous creation of Redwood National Park and ending the threat of dams in Grand Canyon National Park. But others deserved credit, especially the North Cascades Conservation Council, which was on the ground in the Northwest and marshaled those witnesses and kept the newspapers apprised.[62]

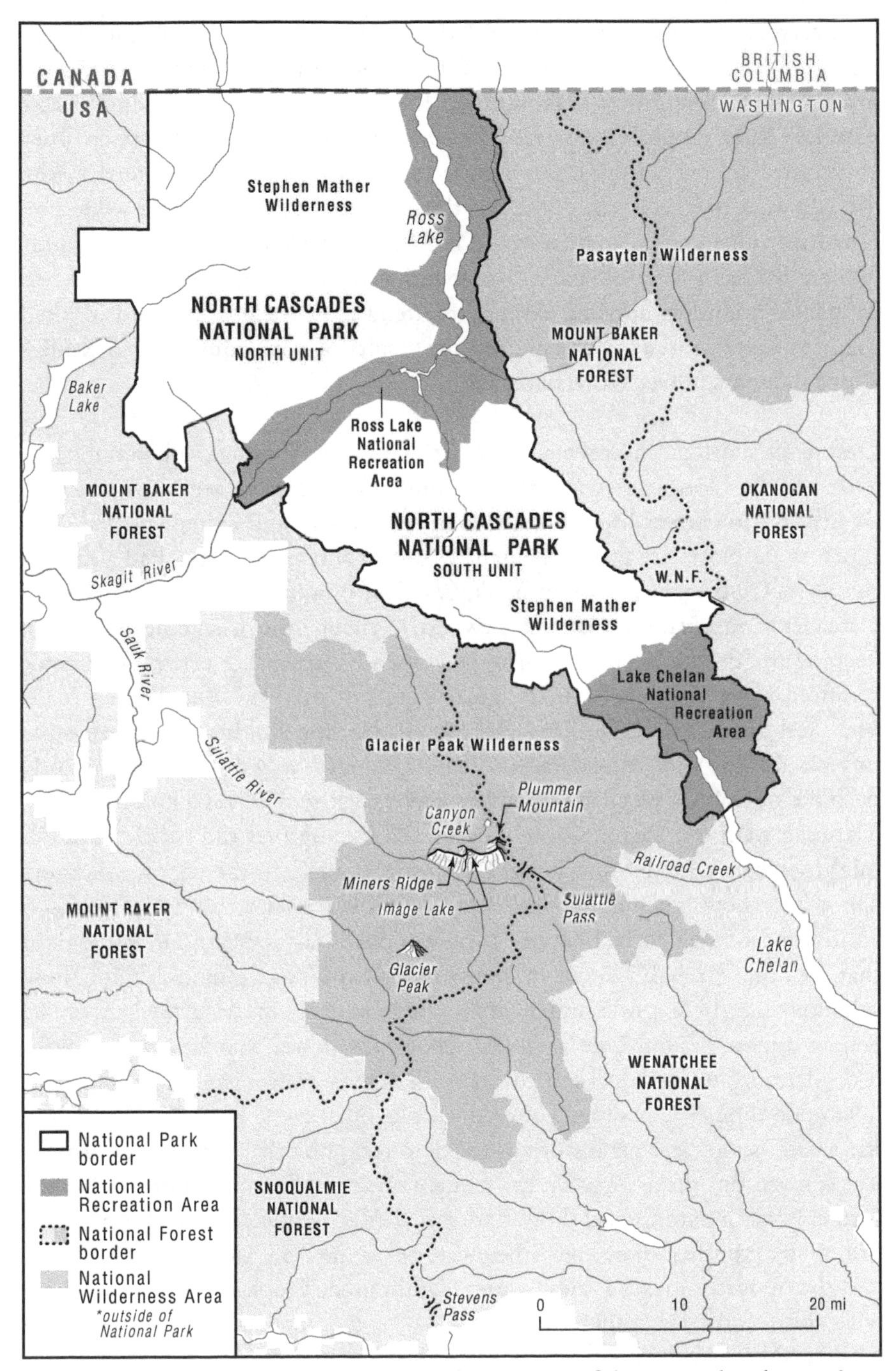

North Cascades administrative boundaries after creation of the national park in 1968

Everyday citizens shaped the legislation—and regional political culture—by repeatedly showing up at hotels for hearings and writing letters to their elected officials. But their power was limited. They rallied to budge the immovable Aspinall. They shook loose some 670,000 acres from the Forest Service. They confronted timber corporations that wanted more timber and hunters who did not want to lose access in the North Cascades. But Glacier Peak wilderness remained vulnerable. Congress would not change the Wilderness Act or include Miners Ridge in the new park. The looming threat of an open-pit mine had motivated countless northwesterners to attend crowded hearings and inspired them to write protests. But laws are the results of compromises, and Miners Ridge did not survive. Work remained.

Despite its shortcomings, when the North Cascades National Park became law, it helped the Northwest conservation community. Jackson sent a pen used to sign the bill to Harvey Manning, the long-time activist and writer who had done much to promote and frame the debate for the public. Manning had authored the Sierra Club Exhibit Format book, *The Wild Cascades,* and edited the North Cascades Conservation Council newsletter, giving him a significant role in keeping the debate alive. Manning had missed the signing ceremony, unable to afford plane fare—"one of the great regrets of my life." But the pen made him "feel much better," he told Jackson. "(Also, my children are enormously impressed.) For your consideration, thank you very, very much. This instantly becomes one of my most treasured possessions—it symbolizes a lot of my life." Manning went on, thanking the senator and hoping that the North Cascades might serve as a model for "integrated area planning" in the region. Manning appreciated how much conservation legislation had passed during the Johnson administration, which in October 1968 was on its last legs. And he recognized that Jackson, "probably more than any other man," was responsible for those achievements. It was not undue praise, because Jackson's role as chair of the Senate Interior Committee gave him enormous power that he was happy to use. Manning wanted Jackson to know that conservationists recognized that. "Platitude though it may be, your full thanks must come from the future, and the record is too clear for historians to fail to recognize that the North Cascades Act is *yours.* But those of us on the scene, now, can offer a small earnest on the future: it was a damn good show, and you were magnificent in a starring role." Manning and other allies knew they were not done. "So, we are all still young, and the opportunities for the future are unlimited; I look forward to working with you toward more such victories," Manning concluded to Jackson. Surely Miners Ridge sat at the top of the priority list.[63]

NINE | The Student

ONE ENDURING IMAGE OF the 1960s is that of the student activist, the politically engaged young person who dedicated time and energy to a cause. Sometimes it is easy to dismiss young idealism, but student activists engaged widely in social problems and movements to solve them, helping to transform American life. Besides the civil rights movement and the free speech movement, the environmental movement brought college students (in this case largely white and middle class) out of the classrooms and off the campuses. By 1969 the Associated Press identified the environment as "a new supercause."[1] College students crafted letters and stood before committees to criticize Kennecott Copper and castigate congressional compromises. Students from the University of Washington formed a critical contingent during the camp-in in summer 1967, and the university's student newspaper routinely included advertisements and letters to the editor opposing Kennecott. But one student stood out in this Glacier Peak campaign, hailing from far beyond the Northwest, suggesting the pull that the North Cascades exerted across the country.

Benjamin A. Shaine arrived in Washington from Oberlin College in Ohio. He described himself once as having a "a love for the great outdoors and a concern for the quality of life that comes from an increasingly industrial society," and he devoted time and energy—intellectual and political—toward the cause.[2] Shaine enjoyed backpacking and climbing, although skiing irked him, once writing from Aspen, Colorado, that "downhill skiing, even here at the best of places, doesn't excite me much. It doesn't have the qualities of awe, quiet, separateness that we find in wandering & wilderness." To him, skiing was merely a sport.[3] Adding to his outdoor interests, Shaine combined dedication and connections to pay dividends. In many ways, he symbolized the cumulative activism for Miners Ridge, and he adopted and adapted the strategies and rhetoric of those who had come before him. Shaine was this era's activism embodied in a single individual.

In the late spring of 1968, just a month after the historic House hearings in Seattle, Shaine followed up a suggestion by Grant McConnell and wrote to Evans, the Sierra Club's representative in the Northwest. As Shaine put it: "I want to do something useful for conservation and the North Cascades and am willing to give my summer to such work." And just to be clear: "Pay is of secondary importance."[4] Evans snapped up the generous offer.

For a college student to spend the summer of 1968 engaged in political action was not news. By this time many African Americans in the South and elsewhere had more than a decade's worth of organizing experience, including 1964's Freedom Summer. Dedicating time to causes helped students like Shaine participate in political change. Students for a Democratic Society, for one organization, focused on participatory democracy as a central tenet. By 1968, the hopes of the early sixties had splintered into many directions and causes, just as the Democratic Party fractured that summer over the Vietnam War during its convention in Chicago. But students like Shaine demonstrated idealism in the face of political practicalities, and they displayed willingness to persist despite a political system that could seem immovable or indifferent. While the era seemed to be unraveling by 1968—with reversals in Vietnam and assassinations in the United States—Shaine brought his energy, intellect, and dedication, though the record does not explain why Shaine chose the North Cascades.[5] Grand Canyon and California Redwoods also faced threats, and the polluted Cuyahoga River, much closer to home, was about to burn, but the American Alps called Shaine, demonstrating how important this cause had become nationally.

Shaine's offer to work with Evans was not uninformed or impulsive. During his junior year at Oberlin and before writing to Evans, Shaine researched the issue. He created a short report that drew on a range of sources from Kennecott, conservationists, writers, and politicians. His purpose was "to generate an informed protest against the Kennecott proposal." The honors student proceeded to describe and assess the situation as he saw it. Shaine emphasized the incompatibility of wilderness and mining as well as the economic insignificance of the mine. As he put it, "The nation does not need the copper deposits of Miners Ridge. . . . But the nation does need the Glacier Peak Wilderness Area." The American people, he wrote, should "use our influence to prevent the development of an open-pit mine in the Glacier Peak Wilderness. And we must insist that our legislators revise the Wilderness Act to exclude all mining and all other inappropriate development in wilderness lands." Shaine's paper contained little that was original in its information or its solutions, but he had persuaded six faculty members—including two department chairs and two full professors—from biology, economics, geology, psychology, and sociology to sign on—to "endorse this report and urge the Kennecott Copper Corporation to abandon plans for mining in the Glacier Peak Wilderness."[6] Shaine meant for his words to move others. And he later called on professors again for practical assistance and to give his words authority.

Shaine offered to work with Evans at a historical moment when young people were influenced by intellectuals and scientists deeply concerned about the environmental crisis. When he introduced himself to Evans, Shaine explained that his "special interest" was how technology affected people, their values, and ecology. This interest coincided with of one strand of emerging environmental awareness, a strand typified by theorist Herbert Marcuse. Historian Robert Gottlieb characterized Marcuse as "a kind of underground intellectual hero to New Left activists." Marcuse's 1964 book *One-Dimensional Man* influenced college students who found its critique of science and technology a useful way to understand the ways industrial society dominated nature and people.[7] The biologist Barry Commoner extended this critique in his popular 1971 book *The Closing Circle: Nature, Man, and Technology*, best known for its four laws of ecology: everything is connected to everything else, everything must go somewhere, nature knows best, and there is no such thing as a free lunch. *The Closing Circle* indicted industrial society's technological choices. Since World War II, Commoner maintained, the economy used technology to produce pollution more than products needed for living, and by doing so it undermined the ecological basis of society. Further, Commoner criticized the U.S. economy as the source of all injustice disproportionately harming poor and other marginalized people.[8] Even as Commoner used the environmental crisis to condemn American economic arrangements, another biologist, Paul Ehrlich, homed in on overpopulation as what he saw as the single problem driving global ecological disaster. His 1968 best-selling *The Population Bomb* introduced Americans to apocalyptic scenarios of starving masses worldwide. Ehrlich's knack for self-promotion and a certain bombast linked population and environmental crisis in the public mind.[9] The debate between Commoner and Ehrlich—simplified as pollution versus population as the root cause of environmental problems—animated many a campus discussion in the late 1960s and early 1970s.[10]

Shaine saw no reason to choose sides. While famous biologists argued over social causes and effects, Shaine—a sociology student—saw these problems as a reason to embrace wilderness for biological reasons. In the report he produced during his junior year, Shaine wrote, "As technology and population alter the environment, there is an ever increasing need to protect ecological units in their original state for study now and in the future."[11] What called the midwesterner to the Northwest, then, was an opportunity to protect one of those ecological units still mostly intact. Although he never developed the ecological argument fully, its appearance here presaged the future of wilderness activism rooted in concerns about biodiversity.[12]

When Shaine first contacted Evans, he was planning to write an honors thesis that would constitute a regional plan for the Pacific Northwest, "an investigation into the best allocation of the resources of the area, the best use of its lands, the best design of its cities." The plan was ambitious—overly so, it turned out. He hoped to gain "contacts, knowledge, and perspective" from his time in the Northwest in the summer of 1968, and "as much time in the wild Cascades as possible." Evans obliged.[13] But Shaine always looked beyond the classroom; his academic and activist work mingled.

That summer Shaine became well acquainted with two Northwest conservation leaders, Evans and Pat Goldsworthy, who led efforts to protect the region's wilderness. One memorable task Shaine shared with Goldsworthy would be a July 1968 trip to Goose Prairie, Washington, to interview William O. Douglas. This conversation followed up from the camp-in. The Sierra Club was making a film about Kennecott's invasion and needed audio recordings of Douglas. The young man from Oberlin surely found a Supreme Court justice a suitable contact for his research. After this immersion in Northwest conservation circles, Shaine returned to Oberlin with a much-deepened sense of the North Cascades and its conservation controversies.[14]

Shaine produced an impressive honor's thesis, long and weighty.[15] He never saw it as strictly academic, though. Writing to Evans, Shaine explained that while his project was "a series of papers on major issues in Northwest conservation," engaging in the research gave him "a lot of time to work on practical activism." In his thesis he spent more time explaining his plan for establishing "a co-ordinated publicity campaign to greet Kennecott when and if they move" than on scholarly conclusions. He believed that his "definitive, scholarly analysis" might be marshaled to inform "the important people." This artifact of the 1960s drips with idealism. He told Evans, "I must know everything there is to know about the issue and its ramifications," and he said, "I'd like to make my report truly definitive." Shaine stood ready to challenge Kennecott and organize students (and faculty) to do the same. He believed scholarship could persuade and should be used in effecting political change. He was translating academic theory into practical action in the best tradition of the engaged scholarship in the era's sociology.[16]

At its academic core, Shaine's thesis considered the supply of wilderness and copper and the ways available for the public to respond to Kennecott's proposal. Although Kennecott and Glacier Peak were the focus, the questions about wilderness, extraction, and the public extended far beyond Miners Ridge, so Shaine's framework transcended place. He asked critical questions: How can we define the public interest and enact it? How can resource allocation, including wilderness, be determined in the United States? When it comes to determining national priorities and allocating scarce resources, what are the limits of corporate power? These questions applied to much more than Miners Ridge. As Shaine put it, the case study was "a good one for the analysis of the relationship of the corporation and the public interest, of the controls exerted by society on the corporation and the powers of the corporation in determining for society how scarce resources will be used." Yet his thesis was deeply rooted in the recent political history of the North Cascades. Central to Shaine's understanding of wilderness was the idea of its being "intact, undisturbed," something that Glacier Peak Wilderness Area provided almost perfectly. The entire wilderness, with the snow-capped volcano at the center like a sun in a solar system, was a unit; any disturbance would wound the "integrity of the area." This integrity meant that Kennecott threatened to violate more than just Miners Ridge, more than the few hundred acres the company proposed to dig up.

The thesis included two parts. The first was mostly a meticulous re-creation of the campaign, Kennecott's thrusts matched by conservationists' parries, with those larger questions of resource allocation fading against the foreground crowded with details.[17] The second analyzed legal strategies available to conservationists and demonstrated not only Shaine's knowledge of law but also a firm sense of conservationists' legal and political strategies.[18] Shaine saw three ways conservationists might use law to stymie Kennecott. The first was county zoning. At the beginning of this campaign, Mt. Baker National Forest supervisor Harold Chriswell explored this option. Miners Ridge was zoned as rural, which precluded industrial activities like open-pit mining. Kennecott required a conditional use permit to proceed, and conservationists could appeal any change that would permit mining. Shaine worried about this option, reading precedent as suggesting that conservationists' standing might not be firm. Secondly, the Forest Service might regulate Kennecott's method of mining or its access. The Forest Service signaled by 1967 that developing Miners Ridge would not be welcomed, but some local foresters felt tied up by the Wilderness Act's allowance for mining. Shaine recognized that any regulatory effort would be "primarily a delaying and harrassing [*sic*] technique"; it would not stop Kennecott. A third possibility was the riskiest but might lead to a permanent solution: challenge Kennecott's constitutional right to harm the environment.[19]

Shaine explored the last strategy with the help of others, especially Charles Wurster and Victor Yannacone, among the cofounders of the then-new Environmental Defense Fund (EDF). Environmental litigation specialists like EDF were just emerging in the aftermath of a 1965 federal circuit court decision in *Scenic Hudson Preservation Conference v. Federal Power Commission*. That decision confirmed that citizens without an economic stake enjoyed standing to sue in environmental cases, and it ordered public agencies—in this case the Federal Power Commission—to consider environmental concerns.[20] Building on this momentum, the EDF approached environmental problems in a way that reflected Yannacone, characterized by Shaine as "an arrogant fellow" and "aggressive" and in scholarly literature as bold, brash, and combative.[21] In this era before the National Environmental Policy Act, which opened more avenues for citizen participation in litigation, Yannacone aimed to create precedents through lawsuits establishing a constitutional right to a clean, unpolluted environment.

He soon gained attention for his Earth Day speech that captured his swagger in its title: "Sue the Bastards."[22]

Because of his larger ambitions, Yannacone viewed Kennecott with his ready-made strategic vision. He wrote supportive letters to Shaine and Evans, himself a lawyer looking for ways to stop Kennecott, but he found their focus on zoning and USFS regulations to be incommensurate with the problem. Yannacone gruffly told Shaine that the zoning approach was "doomed to failure" and might prejudice future legal actions, while relying on the Forest Service "would have much the same effect as the cork in a small boy's pop gun against a charging elphant [*sic*]." Yannacone tried to impress on Shaine and Evans that corporate "lust for land" was unstoppable using "conventional small town legal approaches." He and other EDF leaders dreamed bigger dreams: "To put it rather bluntly, there is either a basic constitutional right to the existence of the Glacier Peak Wilderness as a National Natural Resource for the benefit use [*sic*] and enjoyment of all the people of the United States without degradation as a wilderness ecosystem by Kennecott activities, or there is not." It was that simple.[23]

This constitutional argument appealed to Evans and Shaine, who spelled out the strategy's merits in his thesis.[24] Aiming directly at Kennecott, the perpetrator, surely seemed more satisfying than the indirect, slow process through appeals to county zoning boards or agency decisions. Also, if the strategy succeeded, a constitutional basis for wilderness protection would serve as a powerful precedent. However, Shaine recognized such a decision would be unlikely. Courts would try to find narrower reasons in their judgments, and locating a right to a healthy environment in the Constitution might be a stretch too far for federal judges. To Shaine, the constitutional strategy represented a publicity advantage, though. Rather than appealing regulations on zoning or road construction, a constitutional challenge offered a dramatic way to generate attention. Shaine hoped—and believed—that the public would pressure Kennecott sufficiently for it to end its pursuit of Miners Ridge copper. A growing public opinion in favor of protection against pollution might even sway the courts, Shaine thought, as he recognized shifting legal opinions then happening. All EDF needed was a bit more scientific information, which contributed an essential element to the litigation. Yannacone told Shaine the material the student had already compiled nearly was sufficient to fill out a master brief EDF used for such cases, a mark of the thoroughness of Shaine's research.[25] Yannacone pointed out what Shaine already knew: the plan was risky but offered the possibility of greater success.[26]

At a certain point, Shaine's own actions became part of the thesis. He had come to understand the controversy's history as well as anyone not directly involved, but then he got involved. Shaine developed a petition that urged Kennecott "to permanently refrain from developing its claims." Working with Oberlin biology faculty Edward Kormandy and David Egloff, Shaine distributed the petition nationally and gathered 428 signatures. The notable among them included David Gates, who advised government agencies and Congress and lamented in 1968, "We will go down in history known as an elegant technological society which underwent biological disintegration for lack of ecological understanding."[27] Lawrence Slobodkin and Robert Whittaker, both pioneering ecologists who examined evolutionary systems, also signed the petition. Stanford University was represented by Donald Kennedy, the biology department's chair, and Paul Ehrlich, arguably the most famous scientist in the public eye.[28] The petition surely contained other important scientists who decided to speak publicly on political issues. Not all scientists agreed that their role included public engagement, fearing that some might question their objectivity. Others, Barry Commoner and Rachel Carson prominent among them, argued that various impending crises—from nuclear fallout to poisonous pesticides—demanded a public and critical presence from biologists, especially in a democratic society.[29] Such a response gratified the student activist.

Shaine knew the petition had power but only if used. He had to transform its potential energy to kinetic power. Knowing how to press his causes in high places, in the spring of 1968 he traveled to Washington, D.C., to meet with his own U.S. representative and visit with members of the House Interior Committee.[30] There he almost certainly pushed legislators to support the North Cascades park bill. After the park was enacted, he moved his focus to Kennecott's unfinished business. He wrote to company president Milliken, using the petition as a lever to pry open Kennecott's corporate offices; Milliken agreed to a meeting.[31] Writing to Evans, Shaine hoped that his work might ease "a bit of the burden off your shoulders and, if we are lucky, maybe even get some assurances from Kennecott that they don't plan to go ahead."[32] Shaine proved too optimistic.

On April 11, 1969, Shaine ascended to a "plush executive suite high above Manhattan traffic" and met for forty minutes with Milliken, Kennecott metal division president C. D. Michaelson, the company's director of public relations Edwin Dowell, and its vice president in charge of exploration (whose name was not recorded). Shaine recognized the class of men he confronted: they exhibited "the polite, reasoned, self-assured demeanor of powerful leaders" who considered opposition "insignificant."[33] Shaine hand-delivered the petition and explained to the Kennecott management that the signatures represented a growing number of Americans who opposed actions that destroyed scenic wonders. These scientists and citizens believed "that in our complex, crowded society corporations must accept a responsibility for maintaining the quality of the environment" and insisted that "corporations act in the long-term public interest, even at some financial sacrifice." They were even "willing to forgo a portion of this nation's potential for affluence in exchange for the preservation of the best of our wild and scenic lands free from commercial development." This perspective suggested contemporaneous trends, from the War on Poverty to the counterculture, that asserted or at least suggested the need to stop pursuing maximum profit.[34]

What Shaine presented owed something to the consumer advocacy and corporate responsibility movement spearheaded by Ralph Nader or the efforts of Saul Alinsky. One strategy the Oberlin student deployed, much as Fred Darvill had done before, was to appeal to Kennecott Copper's responsibility as a corporation. Shaine impressed on its "top brass" the disadvantages that would accrue to Kennecott if it proceeded to gut Plummer Mountain.[35] At the time, Nader was helping make "public interest" a significant marker of activism, sparking public interest research groups around the country.[36] Shaine drew on similar rhetoric in the corporate boardroom. The company's reputation would plummet in the face of continuing controversy, he said. Shaine pointed out that the controversy surrounding Kennecott had swirled for two years, and opposition had "not subsided," membership in conservation groups was rising exponentially, newspapers nationwide favored wilderness, and the political process had led to a national park in the region. In short, Shaine pointed out that, regardless of legalities, history's tide pushed strongly against Kennecott's bulkhead. Shaine's final pitch to Milliken looked to the future, suggesting that if Kennecott proceeded with its plans, then Congress would likely enact "very restrictive laws governing mining on public lands."[37] Since natural harmony, solitude, and the other gifts wilderness bestowed fit poorly in any corporate calculus, Shaine tried appealing to the executives' corporate values by using a long-term perspective.

The Kennecott officials responded to Shaine's presentation by defending their position on economic and political terms. The company president said that if the mine could profit, he was duty-bound to develop it. Milliken asserted that the national need for copper—the shortage of copper itself, as well as the balance of payments with foreign trading partners—obliged Kennecott to produce more. Even if a private interest offered to buy out Kennecott's claim, the company might not sell because of this duty to produce. Petitions and protests did not matter. From Milliken's position, the public interest had already been addressed when Congress had passed the Wilderness Act and allowed mining to continue. Shaine later assessed: "This, I believe, is his operating assumption, the one he really believes." The continued agitation by wilderness advocates seemed to Milliken—much as it had to Wayne Aspinall during the Seattle hearings—to be for nothing; the point had been already decided. The company president saw multiple perspectives, however, and contrasted them fairly. Conservationists like Shaine saw the Miners Ridge property "as a small part of copper production and Kennecott views the wilderness disturbed as a small part of the wilderness system." Shaine reported that Milliken had even gone so far as to suggest that if all potential mines in wilderness opened, still sufficient wilderness would exist. When considered in Kennecott's offices, wilderness was in no short supply.[38]

Milliken remained unfazed by Shaine's arguments, telling him that conservationist opposition was "small peanuts." Northwestern protests were mild compared with those the company encountered elsewhere in the world. Displaying gross cultural insensitivity, he added, "At least they don't eat you in the Northwest." Milliken insisted that the "harrasement-legal-publicity [*sic*] aspect" of the conservationists' campaign was "NOT a major factor." As Shaine witnessed firsthand, Milliken was indeed the "scrapper" he was reputed to be and "very willing" to battle conservationists.[39] Nonetheless, Shaine still felt confident. "The Kennecott staff has yet to be educated about the power of conservationists," he boasted to Evans.[40] Beyond the braggadocio, Shaine was on to something, having detected a gap between the words and actions of Kennecott leaders. Despite the company's apparent confidence and its president dismissing opposition and negative publicity, four top-shelf executives had listened to him—"busy men rehearsed and attended a meeting with lowly me," which, he said, "indicates public opinion really does matter" even if these men "do not *think* they consider publicity as very important." Shaine concluded in a letter to Evans, "The issue is not dead."[41] He left the Manhattan office with a better sense of Kennecott's perspective, fueling his campaign and adding to his thesis.

Shaine was nothing if not connected and prepared. In advance of his meeting with Kennecott he had worked with Oberlin College to prepare a press release for distribution.[42] He had written to Evans in November 1968 about plans to coordinate a press release to be ready for Kennecott's move and explained six weeks later how his "editor friend from the *Washington Post* [was] helping with suggestions for getting maximum publicity."[43] "Friend" is not too strong a word, it seems, since after Shaine's meeting the paper published an editorial railing against Kennecott, as it had in 1967 after Orville Freeman came out against the mine. In May 1969 the federal government remained helpless to stop the mine, and the company continued to argue that domestic copper production would ease the nation's "balance of payments deficit," a claim the *Post* characterized as "unadulterated nonsense." Noting the petition Shaine had circulated, the newspaper urged litigation to "halt this unabashed assault on our supposedly preserved wilderness." The college student's petition got Miners Ridge national attention again.[44]

Meanwhile, Shaine built on his momentum. He wrote to Representative Lloyd Meeds, explaining the petition and subsequent meeting. Shaine informed Meeds that Milliken felt "virtually a moral obligation to open the mine" to help fix the country's balance of payments problem. National defense—originally a prominent justification—no longer entered the discussion, an absence attributable perhaps to declining support for the Vietnam War. Shaine wished to report the state of the company's thinking, because Meeds had drafted legislation earlier to stop the mine. Shaine inquired about that legislation, questioning what it entailed and whether publicity might help. Clearly, the student was prompting

Meeds and spurring him to act. It worked. Two weeks later Meeds wrote to the Department of the Interior's chief solicitor with questions concerning the government's power to stymie Kennecott. Since the previous flurry of activity in 1967, the Nixon administration had replaced the Johnson administration's stable of supportive bureaucrats. The Office of the Solicitor (in charge of agency legal services) answered Meeds's questions with discouraging legal interpretations, all of which reduced to this: nothing could be accomplished through administrative regulations, and legislative and litigative solutions were nearly hopeless. Although the response must have discouraged Meeds, Shaine, and their conservationist allies, Shaine helped to ensure that the conversation continued.[45]

The young man from Oberlin touched on much during his years studying and poking Kennecott. He questioned the public interest and thought up new legal strategies. He pushed for accountability and probed scientific rationales for wilderness. He helped keep Kennecott's open-pit mine in the public eye after it faded when North Cascades National Park was created. Shaine was a symbol, a figure representing an issue reaching its zenith.

During his meeting in New York City, after reminding Kennecott leaders of public opposition, Shaine learned much about the company's official perspective. One key insight centered on costs. The position executives presented to Shaine focused exclusively on a simple—even simplistic—economic calculation. They dismissed protests as irrelevant and believed that the public had already spoken through Congress, favoring mining as in the public interest. Kennecott officials only needed to calculate whether the mine would be profitable. Operating costs mattered little in this figuring, since they remained comparatively steady. But the price of copper mattered significantly, and it bounced around like a kite in a constantly shifting breeze. At the time of their meeting, the metal division president, Michaelson, reported the price of copper at fifty cents per pound, which "*definitely WOULD* be profitable to mine," Shaine wrote. The operation's value changed as prices did. Michaelson admitted to estimates "varying 'several thousand percent,'" demonstrating the near-impossibility of making accurate forecasts. Taking the corporation at its word, only economics mattered. This may have told the real story, the real trouble with Miners Ridge.[46]

PART THREE
Resolution

TEN | The Trouble

THE CORPORATE EXECUTIVES WHO made time to meet with the student from Oberlin wrestled with a volatile global industry and experienced upheavals abroad and domestically far more complicated than an undergraduate honors thesis. Kennecott Copper Corporation has not made its files from this era available, so we cannot know executives' strategic thinking in full. Yet the public record shows that as the company faced a public that increasingly opposed its aims for Cascades copper, it also confronted unstable domestic and international copper prices and supply, labor strikes, and federal antitrust attention. All these things complicated the company's decision-making and point toward the broader contexts in which Miners Ridge existed. Those contexts linked Miners Ridge to political developments in Latin America, corporate boardrooms across the United States, and a war in southeast Asia. Miners Ridge was not a discrete place. As we have already seen, it was not separate from the North Cascades National Park, despite not being part of it. It also was not remote from Chilean (or Utahn) copper miners, the war in Vietnam, or Peabody Coal. A key lesson, then, is how specific places remain tied up with many other places and processes.

For conservationists, the context—and the conclusion that flowed from it—was fairly simple: Miners Ridge was in Glacier Peak Wilderness Area and had to remain sacrosanct. Although at times conservationists may have seemed idealistic, even naively so, the archival record shows conservation organizations paying close attention to developments concerning international business, indicating a hardheaded practicality that existed alongside their high-minded sincerity. They came to recognize the centrality of copper prices and government policies connected to these prices and kept track of shifts. And, when possible, they cooperated and supported others, such as organized labor, who created problems for Kennecott. In many ways, the history of Kennecott at this moment was one of trouble, as many forces beyond the corporation's control shook its foundations and complicated its plans. In this way too Kennecott served as symbol for the American copper industry then in a notable decline.[1]

Ben Shaine came to appreciate the role of costs, prices, and profit after his conversations with Kennecott Copper executives. In early 1970 Shaine recapped for North Cascades Conservation Council members how Kennecott saw the economic landscape. Copper was selling for 52 cents per pound, a level deemed profitable. Shaine reminded readers of N3C newsletter *Wild Cascades* that the "mine issue has been dictated by the capital cost/copper price ratio." When Kennecott first acquired the property, startup costs were low, and opening the mine seemed attractive. Costs increased in 1965–66, decreasing the company's interest in mining there. By 1967 the company was forecasting a favorable cost/price ratio, only to have costs rise by the end of that year. With copper prices topping half a dollar a pound in 1969, the mine seemed a certainty. But, as Shaine indicated, long-term company forecasts were "always educated guesses, nothing more."[2]

Copper prices in the 1960s changed constantly. The average annual price of a pound of copper had remained steady after Kennecott Copper grabbed Miners Ridge in the 1950s but sharply increased in the mid-1960s, leading to the proposal to mine. The very period Kennecott aggressively pursued its Miners Ridge plans, 1967–68, corresponded to an especially volatile copper economy. To steady this bumpy economy, federal officials adjusted policy to release copper stockpiles, institute export controls, and offer incentives for new production.[3] Meanwhile, shortages and associated price increases persuaded some copper users to search for substitutes whenever possible.[4] A U.S. Geological Survey report noted that high production in the early 1960s forced voluntary cutbacks that reduced the oversupply that had suppressed prices. However, then the Vietnam War boosted demand. Such are typical vagaries of markets, perhaps. But at the time U.S. copper prices were based on an annually negotiated price. This practice ostensibly provided stability but meant that contracted prices could differ significantly from what a company might receive in an open exchange internationally. According to the USGS report, "During the peak demand period of the Vietnam War, 1964–69, the average London Metal Exchange Ltd. spot price was $0.575 per pound, compared with only $0.38 for the domestic producer price." In other words, selling domestically at the contracted price was a bad deal. So at the very moment Kennecott contemplated developing the Miners Ridge mine, copper prices and the industry as a whole experienced unusual volatility.[5]

The international nature of copper trade exacerbated this instability. Following decolonization and nationalist movements in Africa and Latin America, many U.S. companies lost their holdings abroad during the 1960s and 1970s. In Chile, for instance, critics had long argued that U.S. copper companies,

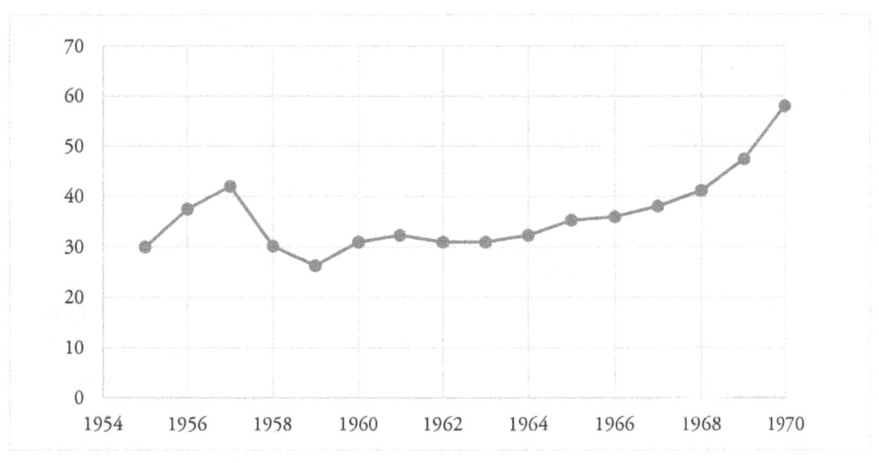

Average annual price of copper (cents per pound), 1955–1970

including Kennecott, profited for U.S. investors but invested too little locally. Kennecott had anticipated the future and in the mid-1960s sold a controlling interest in its major Chilean mine, El Teniente, to the government. Chile experienced political upheaval in the early 1970s. With Salvador Allende's election in 1970, American properties were nationalized. Kennecott lost more than 10 percent of its global production during these years.[6]

If determining prices over the long run challenged the company's prognosticators, costs also clouded the future. In 1967 Kennecott president Frank Milliken outlined some of these costs to Federation of Western Outdoor Clubs (FWOC) president Clark H. Jones, who had written Milliken to urge him to "save" Miners Ridge's "timberline gardens, magnificent vistas, and delicate Image Lake."[7] Although Milliken tried to explain to Jones the basic facts of the Miners Ridge operation and copper mining generally, he was irritated with Jones's organization. An FWOC bulletin had asserted that Kennecott possessed other sites where they might mine copper and thus keep the Cascades wilderness inviolate. Milliken told Jones that this simply was not true, and his explanation revealed much about the broader constraints in copper mining.[8]

Kennecott had spent more than $150 million over five years in the early 1960s to expand its domestic operations and needed copper-yielding property to recoup that investment. Finding copper, Milliken explained, was "increasingly difficult." High-grade ore had been exhausted long before, so the ore that companies located in 1967 ran less than 1 percent copper. "About 15 pounds of copper, on an average, is recovered from each ton of ore processed," Milliken said. But it was even more complicated than that, because for each ton of ore, two tons of "overburden" first needed to be removed. So, for a little more than a dozen pounds of copper, the company had to move three tons of rock. Milliken touted Kennecott as "a leader in recovering copper from wastes which for years were thought to be undesirable," saying it could recover ore bearing less than half a percent of copper because of the company's "significant degree of proficiency of our operations." Would Kennecott go to all this trouble, Milliken asked Jones, if easier alternatives were available?[9]

Still, Kennecott's proficiency could be improved, Milliken thought, with a new experimental strategy. Working with the U.S. Bureau of Mines and the Atomic Energy Commission (AEC), Kennecott sought to develop nuclear energy to aid in their mining operations, an offshoot of the AEC's Project Plowshare program for peaceful uses of nuclear energy. The details that emerged later that fall outlined a significant experiment, a twenty-kiloton nuclear device—roughly equivalent to the bomb dropped on Nagasaki—to be exploded underground near its Safford, Arizona, mine. The device would explode twelve-hundred feet belowground, after which a "solution would be flushed through the broken ore to dissolve the copper and make it available for pumping to the surface." The result of two years planning, the $13 million experiment was meant to test whether this method might be a "useful industrial technique." Officials assured the public that only a "remote" possibility existed of any radiation reaching the atmosphere.[10] The experiment did not happen. But such extreme measures spoke to the growing challenges to the industry, challenges that need to be understood in the context of the corporation's view of the Cascades. (It is not hard to imagine the horror conservationists felt at the prospect of harnessing nuclear power to increase mining's destructiveness.)

Just as conservationists did, Milliken saw the campaign for Miners Ridge as representing a significant obstacle to future copper mining elsewhere, not just in the North Cascades. It was a test. His letter to Jones outlined the company's reasoning and its objections in a way few sources display so well. After explaining the great lengths the company traveled so that it could extract copper, Milliken framed the issue more broadly. Finding ore bodies in 1967 inevitably meant going to places not previously prospected, including "the high mountain country of the West." From Kennecott's perspective, the nation needed "an objective and prudent public land policy" that would balance competing needs of "basic resource development, on the one hand, and our desires for aesthetic and recreational enjoyment, on the other." Milliken was "troubled by the increasing exclusion of prospecting and mining activity on public lands which have minerals potential." A tiny percentage of wilderness would be affected, and he saw no reason why that should be off limits. Milliken worried that if other minerals in wilderness were found, then "each case" would then need to be "tried in the press," a prospect that promised delays as a new (and annoying) cost of doing business. And that business, he implored, operated for the public. "I want to assure you . . ." he wrote, "that we intend to objectively and, I hope wisely, make use of the land we own, and on which we have rights, in the best interest of the nation, our corporation and stockholders, and the public generally." Here Milliken not only reminded Jones of the company's right to develop its claims but also conflated the interests of nation, corporation, and public. What he kept silent about may well have been his desire to ensure essentially unfettered access to development on public lands, something other mining companies argued for contemporaneously.[11] He felt it the company's "responsibility to carry our side of the story to the Congress and the public," which would lead to "[w]ise men" formulating proper public policy.[12]

In his letter to Milliken, Jones praised the nation, saying, "America is the world's greatest nation for many reasons. Our free enterprise system has been a major factor and no dedicated American would change this."[13] But despite Milliken's claim to the contrary, conservationists questioned the presumption that national, corporate, and public interests were one and the same. This discrepancy in worldviews explained much of the conflict between Kennecott and conservationists. But beyond those different perspectives, economic circumstances—what Milliken faced every day—dictated how Kennecott saw Miners Ridge. Milliken showed starkly how much work copper mining took and how much harder it was becoming. Milliken offered this information perhaps to solicit sympathy but at least to explain the complicated calculus Kennecott confronted. Milliken probably saw it differently, but his letter smacks of a desperate industry grasping for all it could find and willing to move six thousand pounds of rock to get fifteen pounds of copper.[14]

So for Kennecott things were hard but soon got worse. As with many industries, copper companies faced strikes by organized labor. During 1966 seven thousand Chilean workers walked out of the Kennecott mines for three months before signing a new contract that raised wages 25 percent.[15] But a domestic strike in 1967 spelled greater problems for Kennecott. The two largest industry unions, the United Steelworkers of America and the International Union of Mine, Mill and Smelter Workers, merged in early 1967, a move that led to "noticeably more contentious" labor relations.[16] The newly bolstered United Steelworkers (USW) sought industry-wide negotiations that focused on wage increases, cost-of-living protections, better pensions, and standardized job classifications. From the USW perspective, the rising profits copper companies brought in—nearly 14 percent net profits in 1966 and 69 percent higher than when the last contract was signed in 1964, by the union's reckoning—matched with the industry's low wages meant that copper workers stood behind other industrial workers and deserved improved compensation.[17] The sides negotiated, and the union even extended the initial strike deadline. Workers continued beyond their contracts, allowing the companies to "stockpile heavily." Finally, on July 15, 1967, Kennecott's workers left their posts; 11,750 workers immediately stopped working. Before it concluded, more than fifty thousand workers participated. As it happened, the USW struck at precisely the time as world copper prices dipped, further allowing companies to withstand any temporary shortages by importing.[18]

The union castigated the "shortsightedness of an arrogant industry" in a pamphlet published two months into the strike. The full details of grievances matter little to this story, but the union sensed that other industries fared better and that improved contracts were also warranted due to copper companies' rising profitability (at "all-time peaks"). The union feared that the federal government might invoke the Taft-Hartley Act and order workers back on the job, something the Steelworkers warned would be a "blind alley." Early in the strike, Kennecott's position was that "the climate was [not] right for settlement . . . and that further negotiations . . . would be a waste of time."[19] A profile in the *New York Times* conveyed Milliken's "reputation as a scrapper."[20] So the strike dragged on. By January 1968, six months into the strike, the Johnson administration appointed a mediation board to bring parties together, an improvisational strategy the *New York Times* said reflected "the inadequacy of Federal labor laws in dealing with disputes in which the concentrated power of giant unions and giant corporations combines to endanger the national interest." The union stuck to its hope for industry-wide negotiation, and this move by Johnson helped move in that direction (just as invoking Taft-Hartley would have helped copper companies).[21] Still, resolution remained distant. To help compensate, the government released copper stockpiles and controlled exports.[22] Not until March did reports indicate that agreements were in place and then only after President Johnson insisted the sides come together again, this time in the capital's Old Executive Office Building, to resolve the strike. In settling it, the union did not gain its industry-wide goals but managed to win economic improvements for its members.[23]

Conservationists paid close attention to Kennecott's business interests, gathering evidence and watching the copper market. A *Wall Street Journal* clipping from mid-1966 remains in the North Cascades Conservation Council records, received in February 1967 as conservationists organized. The article reported on the fading "Great Copper Shortage of 1966," and someone at N3C scrawled, "The truth comes out!" Besides reporting that the shortage appeared to be lessening, the article also explained that many copper users were switching to alternative materials. Similarly, another article from June 1967 noted a surplus anticipated to be one hundred thousand tons by the end of the year, to which someone at N3C noted, "Doesn't sound like the copper shortage is so serious."[24] By the end of summer 1967, with the strike not even two months old, copper executives argued that production needed to ramp up, not because of shortages but to recover rising costs. Once again, conservationists seized on this admission, seeing it as hypocrisy since Kennecott had long said it wanted to develop Miners Ridge to meet national copper needs. Such a reason might seem noble, even to conservationists, but merely to improve business profitability seemed insufficient rationale to endanger wilderness. On a clipping of the article reporting this, in the North Cascades Conservation Council files, is another handwritten comment: "The truth about copper shortages—straight from the horses [*sic*] mouth."[25]

Conservationists made common cause with United Steelworkers. As the strike headed into the fall, Richard L. Miller, the Steelworkers' public relations specialist, contacted Brock Evans to gain information about the North Cascades. It became a fruitful exchange. Evans provided conservation newsletters, newspaper articles, and Kennecott correspondence to fill in the Steelworkers on the conservationists' campaign. He also offered a six-page, single-spaced letter detailing how their opposition to Kennecott had developed; it was one of the most complete records of conservation activity produced at the time. Evans reported: "We feel on a much stronger position than we did a year ago. The public is alerted and aroused, and nearly unanimous in opposing the operation. It is politically popular thing to oppose this, and many politicians are on our side. . . . It is my personal feeling that, while we are not out of the woods yet, this operation will probably not get off the ground."[26] Although hopeful signs appeared, conservationists did not stop.

A key result of this correspondence appeared in the union's newspaper, *Steel Labor*, in January 1968, a story titled "Kennecott vs. the North Cascades." Beneath a beautiful photograph of Image Lake with a ridge of snow-spotted mountains in the background, the story told of a greedy corporation, blind to public interest, pursuing desecration. Kennecott's vice president of mining claimed that the company was responding to a national call for assisting in the war effort, but *Steel Labor* suggested an alternative view: to "[swell] still further the company's already near-record-high profits." In the midst of the ongoing strike, the United Steelworkers now publicly opposed Kennecott's open-pit plan, arguing that the union "better than any organization knows the mockery of Kennecott's expressed concern with the national interest." The article recapped the campaign, rehearsing the arguments it had received from Evans. It concluded with Joseph P. Molony, USW vice president, defending the North Cascades and the public interest: "This seems clearly to be a case in which the interest of the nation and the public must be served, for both our and future generations have a great stake in the preservation of our country's beauty. Foresight and restraint are crucial." Molony noted that the promised jobs "on balance, would be destructive, and this fact must govern our union's position in the matter." The notable statement from a union executive demonstrates one way organized labor often pursued environmental goals alongside their bread-and-butter issues, wages and contracts.[27]

Miller sent an early draft of his article to Evans that concluded differently, showing the importance of public appearances. In September 1967, the American Mining Congress (AMC) met and expressed concern about the industry's image. Representative Wayne Aspinall spoke for many when he characterized copper as an "industry that digs holes, strips vegetation, ruins scenery, pollutes air and water and gets rich quick." Aspinall urged the AMC to "counterbalance that impression," and the organization formed a public relations committee. The union suggested that rather than form PR committees, the industry should "reject past policies which have been inimical to the public interest . . . which have been responsible for the view of the industry held by so many people." The North Cascades supplied a prime opportunity to demonstrate that the industry stood ready to change, and union members stood ready to remind mining executives of their obligations.[28]

Conservationists welcomed organized labor's support. Brock Evans asked the USW for several hundred copies to give to "our union-minded members" to further generate mutual support for wilderness and the strike, adding that Kennecott's treatment of workers—not to mention the land—revealed that the company leaders were not "aware that the nineteenth century [was] over."[29] And the USW was not the only union engaged with Miners Ridge. Senator Jackson heard from the Amalgamated Transit Union, the Window Cleaners Union, and the Culinary Workers and Bartenders Union, all with statements of opposition and disapproval of Kennecott's proposed mining venture.[30] This diversity of union support, even after the USW settled its strike, hinted that broad public support was growing and even crossing class boundaries. Clearly Evans was correct when he asserted that being against the mine had become politically popular for northwesterners.

During this period, Kennecott Copper did not stand still. In early 1968, after years of discussion, it acquired Peabody Coal Company, the nation's second-largest producer of soft coal. The move seemed odd. Kennecott's domestic mines continued to pay, and the company held plentiful cash reserves. But in buying Peabody, Kennecott diversified its corporate holdings rather than devoting that cash to further exploration or improving efficiency. From the initial idea of acquiring Peabody, observers fretted over antitrust implications. Kennecott attorneys believed the deal would satisfy the Federal Trade Commission (FTC), but they were mistaken. In August 1968 the FTC charged Kennecott with restraining trade. Especially in the West, Kennecott would control too much coal, which might unfairly affect utilities. The divestment took longer than the acquisition, still meandering through the courts in 1975 before Kennecott finally sold it off in 1976. This fleeting episode is a reminder of the myriad concerns that Kennecott executives pursued and balanced while a mine site in Washington State awaited their decisions.[31]

The trouble for Kennecott was far from straightforward. The challenges of costs, prices, strikes, and acquisitions disrupted the company's profits. Still, just before the strike, businesses were not hurting much. The first quarter of 1967 saw stock value rise almost 50 percent, from $.70 per share to $1.02, compared with 1966. Both production (up 42 percent) and copper prices (up 16 percent) rose in 1967 to put Kennecott on good standing just before the strike.[32] But the company reported a "drastic" drop in 1967's final quarter. The short-term contrast was stark. In 1967 Kennecott cleared $2.32 per share, just 61 percent of the previous year's $3.82 per share, and the company reported a 16 percent sales drop.[33] During the strike Kennecott took big hits in its profits and relied on compensation from increasing imports from their properties abroad.[34] After the strike ended, Kennecott profitability pressed upward; the first three quarters of 1967 and 1968 amounted to almost identical earnings.[35] In sum: Kennecott recovered from the strike without much difficulty, and, despite slowdowns in profits, the company remained profitable throughout.

Contemporary sources issued mixed messages and reflected rapid changes in the domestic and global copper situation. Kennecott had long claimed domestic need and imbalance in the international market as reasons to pursue Miners Ridge's hidden copper. However, other facts undermined that position. In the summer of 1967, a hundred-thousand-ton world copper surplus was reported just a year after the federal program, citing shortages, called for incentives that pushed Kennecott toward Miners Ridge. By 1969, a domestic shortfall seemed to be rooted in exporting too much copper to take advantage of more favorable prices abroad.[36] When prices and costs rose and fell like so many updrafts, it was hard to plan for a profitable mine.

Through it all, conservationists focused on Kennecott's economic position and did not find the company's shifting arguments compelling. Perhaps nothing would have persuaded Pat Goldsworthy or Brock Evans of the need for Miners Ridge copper. But Kennecott's continued profitability amid disruptive changes served as de facto evidence that the company did not need to dig at Miners Ridge—for the public interest *or* the company's bottom line. The facts transcended all arguments. The military need argument failed. The national interest argument turned out to be inaccurate. Even the business argument looked shaky when profits continued against the disruptive background noise. In the meantime, the grassroots covered the Northwest and beyond with a blanket of support for the wilderness over the mine. Conservationists crafted a superior narrative.

ELEVEN | The Stories

WRITERS OF NATIONAL RENOWN found in Miners Ridge compelling copy. From the conservationists, authors pieced together stories of the North Cascades and placed the region and its controversies into a growing national narrative, constructing it piece by piece in patchwork patterns until Miners Ridge represented a complete quilt. This strategy of constructing compelling narratives was not new. Early in the twentieth century, writers John Muir and Gene Stratton-Porter, among others, sought broader audiences to generate wider enthusiasm for natural history and to protect the local landscapes they cared about. So when authors published articles and essays and even books that included the morass in the Northwest mountains, they followed well-worn literary and political paths, like the long-established Cascade Crest trailbed that skirted the mine site.

When national writers discovered the Northwest, they embedded their stories in larger contexts and meanings. Just as Kennecott plotted its Miners Ridge moves with an eye too cast toward Chile, the London Metal Exchange, or Peabody Coal, writers saw Miners Ridge as part of something larger, the conservation movement that was becoming environmentalism. The heavy lifting for Glacier Peak wilderness always came from northwestern activists like Patrick Goldsworthy, Harvey Manning, and their North Cascades Conservation Council colleagues or the Mountaineers, whose steadfast support and long roots in the Cascades lent them strong regional credibility. But the environmental movement emerged in these years as a national force, especially the Sierra Club that grew under David Brower's leadership. Much as Brower changed the club from a California-based outing group to a national political force, writers changed the Pacific Northwest's latest wilderness bout into a national battle royale.[1] As the campaign for North Cascades National Park captured national attention, profiles of the place and all that was at stake there gathered more than a regional hue, growing to a national glow. The stories of Miners Ridge, Glacier Peak, and the North Cascades—specific regional places—came to represent broader questions about corporate responsibility and the public interest, national landscapes and private profits local jobs and national icons. Talented writers told powerful stories that solidified Miners Ridge's place in the contemporary pantheon of environmental causes. Some of the writing still resonates.

Stories that hit large audiences helped the Cascades' conservation causes. Stories like these helped generate support and enthusiasm that could be marshaled to leverage politicians nationwide. The glossy western magazine *Sunset* had long promoted the region's outdoor amenities. In 1965 it boosted the Pacific Northwest national park cause by publishing "Our Wilderness Alps," a prelude to the sort of coverage soon common in the campaign. The article began with a full-page color photograph of Glacier Peak and Image Lake with long evening shadows, inviting readers into the North Cascades. In the foreground two adults and a child stood on a tiny peninsula jutting into the lake. This high country—towering volcanoes, deep river valleys, tranquil lakes—clearly stood apart. *Sunset* noted, "You can get right up to these mountains by car, but their snowy heartland you reach only by trail."[2] *Sunset* pitched the more restrictive wilderness park option to its upper- and middle-class audience rather than advocate for an automobile-friendly approach. Just a year after its protection under the Wilderness Act, Glacier Peak found itself highlighted by *Sunset* not because of its new status or simply its unquestionable beauty but because of its peril, always an essential ingredient to structuring these stories.

The existing wilderness area, half a million acres, was insufficient, *Sunset* suggested. "People who are alarmed at the rapid loss of natural values in the North Cascades want a substantially larger portion—more than a million acres—to become a national park." The size was too small; the current Forest Service protection was too timid. *Sunset* then guided its readers through other spectacles to see across the North Cascades, to widen the interest in the region and attract a constituency for the park. Although most of the article lauded the region's scenic highlights, its anonymous author did not shy away from pointed remarks. Holden Village—a small religious camp on the east side of the Cascades that occupied the former mining townsite—was "extraordinary," but the now-closed copper mine blighted "the canyon very badly with its mine tailing[s]." New religious occupants could not redeem or reclaim the profaned land. After pages of travel advice, common to *Sunset,* the article weighed in on the political debate about the North Cascades park status Summoning regional history, *Sunset* noted that "logging and mining interests have long and successfully opposed the creation of any national park in the North Cascades." Then it noted that helicopters were landing within a mile of the idyllic scene that had opened the story. Although Kennecott was not mentioned—it had not formally announced its plans to mine there—*Sunset* crafted a story of a place long under siege and warned: "The tide of civilization continues to lap ever higher around the base of Glacier Peak. In fact, as you approach it on the White Chuck River road, through eroding cutover land, you may even wonder whether enough is left to be worth saving as a bona fide natural or wilderness area. Reassurance may come when you stand well inside the remaining wilderness, perhaps on the summit of Glacier Peak itself."[3]

Consider that provocative image: the wilderness advocate, standing atop a glacier-covered volcano, seeing nature extending below in all directions. The vast panorama was shrinking. Clearcuts and roads below her circled and encroached through the valleys. On Glacier Peak, she might be safe for a time. This was what wilderness advocates meant when they referred to rock and ice wilderness, wilderness ever shrinking, ever higher, ever more remote. Like the encirclement the Soviets feared in the Cold War, *Sunset* showed that something needed to be done to preserve the freedom wilderness provided while some wilderness remained.

Sunset was not alone. As Kennecott made its move and as North Cascades National Park was slowly born, more stories emerged, more narratives wound through the mountain peaks, and more authors of national stature took their stands. Paul Brooks exemplified these trends. A member of the Sierra Club's board of directors, Brooks worked as editor-in-chief at Houghton Mifflin, where he edited Rachel Carson to produce some of the finest environmental writing ever published.[4] He wrote a pair of articles in 1967 that were published in two of the nation's finest periodicals, the *Atlantic Monthly* and *Harper's,* widely read by influential elites. Brooks weighed in on the swirling debate about "America's Alps." Brooks likened the rising environmental movement to a second American Revolution, albeit a "bloodless revolt" this time. He urged defenders of wilderness to be loud, and he argued that concern for the environment represented "no less fundamental [a question] than concern for economic prosperity, or for political and human rights."[5] Brooks identified the campaigns centered on the North Cascades as representing national frameworks, explaining the high stakes and arguing to a national audience that wilderness was every bit as much about democracy as it was about scenery.

In "The Fight for America's Alps" in *Atlantic Monthly,* Brooks combined a history lesson of the feud between the U.S. Forest Service and National Park Service with a personal account of his family's trips through the North Cascades. He made clear the superlative beauty of the area, a paradise under siege. From his Boston perch, Brooks recognized better than most the larger significance of what was playing out in the North Cascades. This place—"the wildest, most rugged, and probably most beautiful mountains in the United States south of Alaska"—faced threats because its inaccessibility was diminishing. It was no longer simply a bureaucratic wrestling match of resources users, represented by the USFS, against the recreationists represented by the NPS. Administrative traditions and philosophical visions battled. As Brooks put it, "We are watching the confrontation of rock-bottom conservation philosophies, involving basic principles as well as power politics." But this depiction, with its combatants easily stereotyped, oversimplified things.[6]

Brooks recognized that the Forest Service and Park Service each confronted problems, long-standing and new. Like many (including Grant McConnell, whom he cited), Brooks found the Forest Service too beholden to special interests, namely timber companies.[7] But Brooks also chastised the Park Service for excessive development, from Mission 66, a massive ten-year, billion-dollar project that ended in 1966 and added buildings, roads, campgrounds, and parking lots throughout the national park system.[8] The North Cascades offered each agency an opportunity to bolster or clarify its wilderness credentials. Here, then, they fought a proxy war to purify their images so that they might seem as pristine as a North Cascades alpine lake.

A few years later, Brooks gathered this article and other writing in *The Pursuit of Wilderness.* He edited the 1967 piece lightly and included an update because the national park had then been created. In the update Brooks argued that excluding Glacier Peak from the national park constituted a "major omission." His conclusion was twofold: "The story of the North Cascades teaches us once more that in conservation battles there are no final victories." And "there is no better place than the North Cascades for believers in our park system to make their voices heard."[9] Since the Forest Service and Park Service were still devising management plans for the park complex and larger region of public lands, Brooks's words here might best be described as aspirational.

Brooks wrote the original 1967 article in the midst of the campaign to create the park and the revision amid the efforts to stymie Kennecott. These were of a piece—a sustained, strategic effort to influence policy. Words mattered in a democracy, he believed.[10] And he hoped his words might serve the cause. He corresponded with Scoop Jackson, telling the senator he was planning a book on creating national parks, a book that would highlight the North Cascades. Jackson jumped to cooperate, even at one point saying he would get Brooks galley proofs of public hearing transcripts before they were published. Jackson was pleased that someone "well qualified" was writing about efforts underway. As Brooks worked up his next article, for *Harper's,* he shared a draft with the senator, noting that the theme zeroed in on "a corporation's responsibility to the public" more than on the details of Kennecott's operation. That is, he focused on principles, much as his earlier article contrasted bureaucratic visions. Brooks made perfectly clear that his literary work aspired to be political; when he informed the senator that *Harper's* had accepted the article, he told Jackson, "Let's hope it has some effect."[11]

In *Harper's,* Brooks shared the narrative that conservationists had developed. While his first article contrasted NPS and USFS approaches, this one worked to embed corporations in the public sphere. Citing John Kenneth Galbraith, perhaps the nation's best-known economist, Brooks explained that modern corporations kept growing to the extent that distinctions between private and public enterprises became too entangled to be meaningful. Some corporate entities were no longer distinctly private, separate from the body politic. In 1967 Galbraith had published *The New Industrial State,* in which he argued that modern corporations sought broad goals, a "whole complex of organizational interests," with profit being only one of them. From Galbraith and Brooks's perspective, this expanding e power meant corporations like Kennecott were obliged to "serve the public interest." From that start, Brooks delved into the battle of Miners Ridge to illuminate the point. Brooks knew that corporate "self-denial" was too much to ask for, but he hoped that stockholders might exert pressure and that Kennecott might pull back.[12]

In the end, Brooks raised some interesting possibilities about resolving the standoff and interpreting the conflict at Miners Ridge, while helping them gain wider currency. For example, he indicated that the Forest Service chief wanted the legislative authority to buy up claims. Further, Brooks hinted that perhaps all Kennecott really wanted was to "unload its mining claims on the government." All its threats to the wilderness might be "simply a move toward a profitable settlement." Or perhaps the company sought precedent to mine in wilderness as the Wilderness Act allowed but that no company had tested. Either stood as a possibility in Brooks's mind. Finally, Brooks wondered, why did modern corporations not have to account for natural beauty in their calculations? Such questions began to percolate in the margins of economics, resulting by the 1980s in the field of ecological economics.[13]

In these ways, Brooks framed the contested Cascades provocatively and worked to reveal multiple ways that Kennecott seemed to seek profit, even directly from public coffers. Implicitly and explicitly, he raised critical questions about corporate responsibility and the blurring line between public and private enterprises in modern America. By raising his criticism to questions of basic principle—what right did a corporation have to befoul public resources? what right did corporations have to presume separation from the public interest?—Brooks both included and sidestepped Kennecott. The copper corporation was now further stigmatized, but its plans might also be used as a general critique of bad corporate citizenship. Others took that further.

Not all stories were found in magazines as respectably mainstream as *Harper's.* The 1960s gave rise to new forms of journalism along with student activism, and one New Left outlet for this reporting was *Ramparts,* a magazine published in that counterculture mecca San Francisco. During its late 1960s heyday, *Ramparts* reached a quarter of a million American subscribers, offering influential muckraking and irreverent journalism. *Ramparts* writers ranged from Susan

Sontag and Kurt Vonnegut to Noam Chomsky and Eldridge Cleaver, and the magazine was on its way to becoming what one study claimed "the premier leftist publication of its era."[14]

One *Ramparts* staff writer, Gene Marine, extended his reporting into a book on environmental problems, his 1969 *America the Raped: The Engineering Mentality and the Devastation of a Continent.* A combination of exposé and call to arms—very much a *Ramparts*-styled genre—the book traveled the continent and reported on one environmental threat after another. Marine targeted not just problems caused by wilderness invaders but also highlighted industrial harms to air, urban decay, and more.[15] Through Marine, *Ramparts,* representing the broader and even sometimes fringe New Left, brought Glacier Peak into conversation with activists and causes outside the emerging environmental movement. This new attention served crucial functions, because most accounts of the North Cascades focused on and within the conservation community, a group overwhelmingly middle-class, white, and already interested in backpacking in wilderness. Marine brought young radicals into the fold, as the audience for *Ramparts* encountered Black Panthers, read Che Guevara, and found the Central Intelligence Agency uncovered in the magazine's pages.[16] Obviously, this approach differed from the Sierra Club's rather conservative standards.

Because Marine was no wilderness aesthete in the John Muir tradition, he did not make pristine conditions the sole criterion by which wilderness (and deviations from it) were judged. At one point he was explicit: "We can have a road for spoiled city dwellers or people in a hurry (like me) and still have plenty of space in the Great Cascades for backpackers." The comment might have troubled wilderness purists, a group Marine criticized for being "snobbish" and aimed at backpackers only.[17] Marine and the New Left examined wilderness from a more pragmatic angle. He sought middle ground, a way to accommodate use and recreation.

In Marine's mind, what Kennecott proposed differed from merely logging or building an access road through wilderness; it violated the Cascades' ecology wholesale. The importance of wilderness, Marine concluded, lay in the need to maintain ecosystems for them to function and for scientific study, not for aesthetic reasons.[18] These ideas have become a matter of course today but remained a muted or subordinate part of late 1960s wilderness arguments.[19] In his conclusion, Marine recognized how underdeveloped the ecological arguments for wilderness were (although he probably overstated the case), and this put Miners Ridge into a context distinct from *Harper's* or similar publications.

For Marine, fitting into the *Ramparts* mold required that he make Glacier Peak mean more than wilderness for backpackers. So, Kennecott in the Great Cascades, as he dubbed the region, stood in for a larger project, which Marine called the "engineering mentality." His venomous prose used the metaphor of seduction and rape. Old-time extractors left devastation in their wake, Marine recounted, dubbing them "old rapists." In the 1960s they were "substituting seduction for rape wherever possible but no less determined to have their way with the land."[20] The dynamics had changed. The conservation movement had grown up to counter what some at the time called "land skinners," those bent on resource extraction, quick profit, and moving on.[21] The federal government enacted a series of reforms to limit private destruction, a way to keep "the old rapists in check," as Marine put it. New laws slowed rampant destruction, and new bureaucracies managed natural resources with an eye toward the long term. These efforts did not always work. The "old rapists" found collaborators, often within the very public agencies meant to protect nature, an alignment of industry and government that wreaked havoc on public resources. These allies are called "Engineers," Marine argued, always capitalizing the word. The heavy hand of the Engineer took a heavy hand to the land. To be sure, Marine knew that not all engineers destroyed ecological systems, but he emphasized that a *mentality* pervaded the profession, as well as some other professions (e.g., public relations), that characteristically attacked complex problems by a straightforward method: "It is the simple, supposedly pragmatic approach of taking the problem as given, ignoring or ruthlessly excluding questions of side effects, working out 'solutions' that meet only the simplest definitions of the problem," explained Marine. "It is an approach that never seeks out a larger context, that resents the raising of issues it regards as extraneous to the engineering problem involved."[22] Armed with this mentality, the Engineer divided the natural world as useful or useless, nothing more or less. Such a mentality undergirded Kennecott Copper Corporation, allowing it to justify violating a wilderness sanctuary for a small amount of copper.

In his chapter "You Go through Sedro Woolley"—Sedro-Woolley is a small timber town in the Cascade foothills—Marine aimed his barbs squarely at Kennecott, applying the general principles of the book to this specific manifestation of the engineering mentality. To tell the story, Marine relied on Northwest conservationists like Brock Evans and Patrick Goldsworthy, who shared their experiences and extensive correspondence with the radical writer.[23] That meant the basics of the story followed the conservationist line. So Marine told of the Wilderness Act's "sneaky" mining loophole and described the "monstrously ugly scar" the open pit would leave.[24] Importantly, he also addressed scale, an area where the corporation and activists saw things in distinct terms. Kennecott liked to emphasize how small the mine would be, especially when considered as part of a nearly half-million-acre wilderness. But Marine pointed out the myriad problems with pegging the scale to simply the size of the pit. People would see the "ugly scar" from far away. People would hear the blasting from miles away. Roads and "one hell of a big processing mill" would extend beyond the mine site. But perhaps the most important scale Marine highlighted was the amount of copper Kennecott would gain. According to Marine's calculations, the mining process would create "4.5 million tons of junk" for two days' worth of copper. So ludicrous did the scenario seem, Marine likened it to a laughably improbable Al Capp comic strip.[25]

The story Marine spun and the book in which he unleashed it showed that a broader audience—younger, more radical, less local—found Miners Ridge of interest. For the northwestern peaks and ridges Marine constructed a familiar narrative arc but with an edge. A majestic place with an existential threat was no new story. But Marine's harsh indictments of politicians too timid to confront corporate designs challenged conservation's sometimes staid image. He shared the basic position Brooks had developed but removed the sheen of respectability. Even as Marine framed his story around mentalities and absurdities, another arc emerged, a personal one with larger and lasting revelatory power.

Undoubtedly, most people who learned of this battle royale after the fact learned it not from Gene Marine or Paul Brooks but from John McPhee, a writer for the *New Yorker* who profiled the Sierra Club's David Brower in *Encounters with the Archdruid,* a book that became an instant classic and that still finds fans and prized positions on course reading lists almost half a century after publication.[26] The book drew from articles originally appearing in the *New Yorker* and became a standard account of Brower, earning him the lasting sobriquet "Archdruid."[27] A skilled writer of literary nonfiction somewhat within the New Journalism vein popularized by Tom Wolfe and Hunter S. Thompson, McPhee crafted his essays with the care of a novelist. *Encounters with the Archdruid* became one of his most notable successes, and it introduced a wide, literary public to Brower in an indelible way. McPhee followed along as Brower and three antagonists debated the merits of conservation and development. Brower wrangled with Charles Fraser, coastal resort developer in the Southeast, and with Floyd Dominy, head of the Bureau of Reclamation and advocate for dams in the Grand Canyon. The book opened, though, with McPhee tagging along on a hike through Glacier Peak Wilderness Area with Brower and Charles Park, a geologist and mining engineer.[28] Just as Marine had done, McPhee used Glacier Peak as synecdoche, with Brower standing in for all conservationists and Park for all would-be resource extractors. But unlike Marine's vehement arguing, McPhee's prose opened a space for contemplating Brower and Park, wilderness and mining, by sharing their words and allowing readers to ruminate on their meaning for the future.

McPhee began at a twelve-foot-square shed sitting incongruously high up in the Glacier Peak Wilderness Area, where a small group of hikers planned to bed down for the night. Although surrounded by wilderness, a place with few uses permitted, the shed was used by the Chelan County Snow Survey when it sent people to the mountains to measure snowpack. It also sat within the general claim of Kennecott Copper. Here the Wilderness Act mining exception was manifest, no longer pages of legislative small print but a real-world example of what compromise looked like. The party, according to McPhee, "wanted to have a look at the region while it was still pristine." Park, the most talkative of the party and least likely to care about claims concerning anything pristine, represented mining interests (although he was not affiliated with Kennecott). McPhee characterized Park by introducing the Stanford-based mining partisan as believing "that if copper were to be found under the White House, the White House should be moved." The hyperbole signaled a true believer. Not that Park hated wilderness; in fact, he studied mining because mines develop in "out-of-the-way places."[29] This sort of affection for remote settings sat uneasily alongside a penchant for extraction, but it has been a common one among natural resource professionals and workers.[30]

Park's position put humans in charge, managing directors of nature. "You can't avoid change," Park said. "You can direct it, but you can't avoid it."[31] This perspective found full explanation in Park's book that appeared in 1968, *Affluence in Jeopardy: Minerals and the Political Economy.* The book explored his theme: the need to maintain affluence by developing all economic resources possible in as responsible a way as could be devised. Development was not optional. Minerals created the foundation of modern society, in Park's view, and shortages or lack of access might destroy "modern civilization as we know it." No small stakes. He thought smart people misunderstood these basic conditions of modernity and urged them to wrestle with reality: the impending shortages. In doing so, they would be forced to agree that the very foundation of the good life was at stake. Affluence generated freedom, Park asserted in his conclusion; without affluence from mineral wealth, poverty and starvation invariably followed.[32] This perspective guided Park in the mountains. "My idea of conservation is maximum use," he explained to McPhee. "I think preserving wilderness as wilderness is a terrible mistake."[33] This was one character in the drama unfolded by McPhee.

Brower offered something else entirely. Whereas Park seemed gregarious and even bombastic, Brower crept through the pages, reminding everyone on the trail that they were intruders there. While Park hammered seemingly every passing rock with his pick, Brower ambled up the trail plucking random thimbleberries to fuel his ascent. As Park imagined immediate mineral needs, Brower mused with pleasure that a downed Douglas fir would rot as part of timeless natural cycles. The contrasts could hardly have been starker.[34]

With the characters introduced, McPhee moved to setting. He could craft scenes as well as any writer, but characteristically McPhee let a friend, a Park Service employee, speak for him: "The Glacier Peak Wilderness is probably the most beautiful piece of country we've got. Mining copper there would be like hitting a pretty girl in the face with a shovel. It would be like strip-mining the Garden of Eden." If that image of incongruity, of violation and violence and desecration, did not suffice, McPhee shared how each member of the party reacted when crossing Cloudy Pass and seeing, for the first time, Glacier Peak's immensity:

"Wow." "My God, look at that." And silent awe. Brower called it the "scenic climax." Ever the contrarian—the role for which he was cast in this drama—Park demurred: "A hole in the ground will not materially hurt this scenery."[35] Such repartee must have made for a lively hike. Back and forth this conversation went, McPhee listening in on the antagonists and reporting the ongoing score all the while creating an unforgettable portrait of place and people and controversy.

On one side, Park spokes for copper, for multiple use, for humans directing change. On the other, Brower stood for something simpler: balance. "I'm trying to do anything I can to get man back into balance with the environment," he told McPhee by way of explaining why he tirelessly traveled the country speaking out, organizing, and haranguing his fellow citizens. Park was not insensitive to beauty. "I would prefer the mountain as it is," he explained before adding, "but the copper is there." Brower spokes of obligations and higher planes: "Copper is not a transcendent value here." Park expected economic growth and praised affluence, as Brower supported lowering the standard of living. Where Park stood for the present and was unwilling to sacrifice it for the future, Brower worked for tomorrow and was willing to allow future citizens to decide. Park thoughts you could not ruin mountains; Brower knew you could.[36]

These personalities, perspectives, and principles clashed and contrasted, sitting side by side as uncomfortably as an open pit mine in a wilderness. On the hike, each man faced evidence that undermined his view. Early on, before climbing up high, the group passed through Holden near the majestic glacial Lake Chelan where, beginning in 1938, the Howe Sound Mining Company mined for two decades. In its wake, messes remain. A "god-awful mess" was how Brower described it.[37] McPhee painted the scene in part:

> What remained of the mill itself was a macabre skeleton of bent, twisted, rusted beams. Wooden buildings and sheds were rotting and gradually collapsing. The area was bestrewn with huge flakes of corrugated iron, rusted rails, rusted ore carts, old barrels. Although there was no way for an automobile to get to Holden except by barge up Lake Chelan and then on a dirt road to the village, we saw there a high pile of gutted and rusted automobiles, which themselves had originally been rock in the earth, and, in the end, in Holden, were crumbling slowly back into the ground.

McPhee's description damned without him saying so. Prosperity's window closed quickly in Holden. The mine only operated for two decades, a meager economic boost. A single decade after being abandoned, the mine's industrial detritus now produced nothing but rotting, rusting buildings that mar an otherwise stunning "backdrop of snowy peaks."[38] The abandoned autos epitomized the short life cycle of industrial design and consumer products. This was Brower's point. A mine's productivity lasted a short time; its damage lasted a long time. Its products were transitory and soon reclaimed, incompletely, by the earth. For Brower, no matter how you calculated it, mining wilderness was not worth it.

Park could not disagree: "Holden is the sort of place that gave mining a bad name. This has been happening in the West for the past hundred years, but it doesn't have to happen. . . . Traditionally, when mining companies finished in a place they just walked off. Responsible groups are not going to do that anymore." Listening to Park would lead you to believe that the Holden mine represented bad practices from an age long past. But Howe Sound Mining Company had stopped operations just a decade before; contrary to Park's assertion, Holden offered an object lesson in contemporary mining practices. Park suggested, "Grass could be planted on the dump and the tailings."[39] Cosmetic changes only. Today, almost six decades after Howe Sound skedaddled, Rio Tinto remains in the area cleaning up contamination.[40] As Brower pointed out, the only way to avoid such problems was to avoid mining.[41]

But if trailside evidence flawed Park's case, Brower's was not without its own on-the-ground challenges. Many a hiker stopped at Image Lake, the small alpine tarn named for the almost-perfect reflection of Glacier Peak it offers on clear days. The popular destination drew Seattleites for weekend outings, while long-distance hikers paused there along their grueling treks over the Cascades or along the crest. A popular hiking guide published in the 1980s acknowledged, "Solitude is not the name of the game here."[42] It was no different in the late 1960s. When McPhee's "somewhat bizarre group" settled into the Image Lake basin for the night, they struggled to find a place to camp. "We walked past tents along the shore—blue tents, green tents, red tents, orange tents," McPhee cataloged. Besides the colorful lodgings, the group faced noise: "The evening air was so still that we could hear voices all around the lake. We heard transistor radios." Working to get away, the writer climbed to watch the lake reflect the mountain in the day's waning light. "Objectively, the reflection was all it was said to be. But a 'No Vacancy' sign seemed to hang in the air over the lake," McPhee lamented, in one of his few direct interventions. Brower founds the setting wild, wished for more solitude, but took the long view. "All of these people certainly diminish the wilderness experience," he told McPhee, "but I've seen crowds in wilderness before. I know that they'll go away, and when they go they haven't really left anything."[43] The Sierra Club leader spun the crowds in a good light, but clearly he found the congestion troubling too.

Park pounced at the crowds and then launched into a broader attack. "This is no wilderness to me," Park complained. "There's just too many people." Here he continued his frequent commentary about overpopulation, then a popular cause and debate in environmental circles (and an area with which he found common ground with Brower).[44] Park was just warming up, though. The crowds bothered him, to be certain, but now as the party circled Image Lake, the engineer began deepening his second thoughts. "I don't really understand why they come here, though," he said. "This is a very ordinary little mountain lake." (Dismissing a jewel like Image Lake was sacrilegious to those who considered it the scenic climax of the North Cascades.) He continued: "The more I see of this country, the more I fail to see what that copper mine would do to it. When we started, I was under the impression it might do something, but, golly, I can't see that now." The conclusion Park reached toward—that this place was not wild, not special, and a mine would not harm it—derived from his firsthand engagement with the landscape, an encounter that underwhelmed him. The group separated for the night, Brower moving to higher ground, Park staying low, grumbling as he drifted off: "I know half a dozen lakes like this that I can drive to and where I would find less people. I'll give you my interest in Image Lake for a piece of a counterfeit penny."[45]

Their standards, of course, were mismatched. Park's environment served human needs. Brower's did not. Park's resources were endlessly exploitable. Brower's had limits. Park's nature was adaptable and expansive. Brower's was vulnerable and threatened. "All a conservation group can do is to defer something," Brower believed. "There's no such thing as a permanent victory. After we win a battle, the wilderness is still there, and still vulnerable. When a conservation group *loses* a battle, the wilderness is dead." Park disagreed. He thought it a shame more people could not see Glacier Peak; they had the right to, and getting there on a mining road might make the difference. Brower believed that such a view "ought properly . . . [to] be earned . . . on foot."[46] The two viewed the wild, the mountain, differently at the very foundation.

The next morning, the group completed its Cascades traverse, trudged down the mountain, and pointed weary toes toward the Suiattle River, "a headlong chaos of standing waves and swirling eddies, white with spray and glacial flour," according to McPhee. For decades, the heavily-timbered Suiattle drainage drew the attention of timber companies like moths to a flame for its enormous Douglas firs. Just as faithfully, wilderness advocates swatted back loggers' attempts to profit from board feet extracted from trees hugging the river. Evidence of Forest Service improvements lay on the forest floor—rocks dynamited apart, trees knocked down, a wider trail, an easier slope. Brower began complaining when the group encountered the USFS employee in charge, a man with enough disdain to shed on "wilderness-lovers" and miners alike. At trail's end, just before McPhee rushed to plunge aching feet in the icy glacial melt of the Suiattle, a mechanical rumbling sounded, louder than the rushing river. A bulldozer. "Half submerged, its purpose obscure, it heaved, belched, backed, shoved, and lurched around on the bottom of the Suiattle as if the water were not there."[47] McPhee left the machine in the river without commentary—none from his own thoughts, Brower's, or Park's—just the invasive machine in the wilderness, chugging.

As character sketches of Brower and Park, *Encounters with the Archdruid* left an impression. For the reading public of the *New Yorker* who were more comfortable taking a bus across Central Park to the American Museum of Natural History than the causes of conservationist circles, this portrait might have been their only view of Brower, and, as such, it widened consciousness of the cause just after the first Earth Day. Park, a professor of mining engineering, probably never entered public discourse on such a scale again. Although he functioned well as a sparring partner and gave a reliable quote, Park ultimately proved interchangeable—he had been a last-minute substitute for a Kennecott executive—because anyone from a mining company (or a logging company, for that matter) might have stood in to debate the man billed as the "archdruid." The story here, told through a debate conducted along a trail that climbed over a major wilderness mountain range, amounted to a morality play. McPhee picked no sides; he observed, listened, and shared the arguments—something perhaps no writer has done better.

As Brower jousted among the ridges and valleys, his large significance rose like Glacier Peak itself. He was embattled at the Sierra Club he had rebuilt. Being executive director did not make him Sierra Club dictator, as photographer Ansel Adams had once derided him. Soon Brower had to answer for his impetuous style, especially with finances. Brower's bold leadership built the club into a political force, and longtime club leaders recognized this role he had played and valued him. Yet Brower spent money without approval; he launched campaigns without board blessings; he contradicted club policy; he issued advertisements deemed political that caused the club to lose its tax-exempt status. These problems flooded, pouring over the levees his allies built in a last-ditch effort to contain his unruliness. In September 1968, the month before North Cascades National Park was born, some Sierra Club board members sought unsuccessfully to oust Brower. The next year, in May 1969, as Ben Shaine parried with Kennecott executives, the board succeeded in voting him out. Brower resigned first; the directors accepted it on a 10–5 vote. Brower told the membership, "We cannot be dilettante and lily-white in our work. Nice Nelly will never make it." He had, as one biographer put it, "outgrown the Sierra Club."[48]

McPhee was present for the ouster in San Francisco's Sir Francis Drake Hotel. In the final section of *Encounters with the Archdruid,* McPhee tallied votes and profiled directors in spare sentences, concluding each with either "Pro-Brower" or "Anti-Brower." One of the directors who no longer supported Brower was Paul Brooks, his specific reasoning lost to time.[49] Brooks's presence in the hotel's fittingly named Empire Room, along the ubiquitous U-shaped table meant for boards, brought this story of stories to a close in a neat package. Brooks had joined the North Cascades cause prominently, writing in *Harper's* and *Atlantic Monthly.* With McPhee's *New Yorker* pieces followed by a stand-alone book as a coda, the media made Glacier Peak and Miners Ridge something nationally notable. None of that meant that all advocates advocated for the same reason, as

Brooks's sliding support for Brower suggested. But dissonance in the Sierra Club also did not indicate faltering commitments to keeping Miners Ridge whole, not holed. Writers told the Miners Ridge story in myriad ways—from celebration and promotions to partisan broadsides—but with McPhee's nimble mind and pen, all sides were laid bare; the debaters debated out in the open, along the ridges, and in the woods. Rather than steeping in partisanship, McPhee's narrative offered something more objective by presenting Park's and Brower's views in their own words. Stories shape understanding, and McPhee gave the reading public an opportunity to decide for themselves.

Sunset and Brooks, Marine, and McPhee all accomplished and symbolized several things. On one level, they brought the North Cascades cause to the American reading public. Since public outrage could motivate political action and dislodge reticent legislators, media attention led to increased mobilization against Kennecott Copper Corporation. In this way, these writings continued the campaign that local conservationists had begun, going back to efforts to protect Glacier Peak wilderness before the Forest Service and later the Wilderness Act granted some sanctity. Local conservationists—from Polly Dyer and Grant McConnell in the early years to Patrick Goldsworthy and Brock Evans later—remained firmly rooted in the Northwest. They lived there. Their organizations, for the most part, drew inspiration from and remained devoted to specific regional landscapes without dissipating their energies much on broader questions and places. The North Cascades Conservation Council embodied this local rootedness. When Paul Brooks wrote about the North Cascades in *Harper's* and *Ramparts* writer Gene Marine presented a parable of corporate greed to the New Left, they incorporated the local into the national.

Nothing represented that better than David Brower. When his personal story intersected with that of Miners Ridge in McPhee's prose, and with his near-simultaneous ouster from the Sierra Club, Brower represented a convergence of the local and regional with the national, of the perils of smaller battles being subsumed into larger ones. Brower's fall represented nothing so much as the growing national (even global) urgency Brower felt and that his acolytes prepared to carry forward to all corners, as it veered off from the old guard in the Sierra Club who still remembered fondly the days of summer outings in the Sierras before politics gripped the club too tightly. Brower left the club to form Friends of the Earth, making this break—or transcendence—obvious.[650] More than ever, the global connections of Miners Ridge became manifest. And copper in the wilderness in Washington seemed not to be separate from emerging global crises. It was not just this mine, in this place, at this time, but part of a much larger environmental cause. The story of Miners Ridge, —of the North Cascades and wilderness against despoilers—took on hues shaded by all these parts, moving and transitioning as the sixties turned into the seventies.

Each author told stories. The public read these stories, which some people discovered resonated with their own experiences. For others, the stories (and photos) would be all they would ever know of the North Cascades. But stories inspired and mobilized, whether readers would ever climb Miners Ridge or see Glacier Peak reflected in Image Lake. The stories were warnings, cautionary tales of mines gone awry, of wilderness overturned like Dust Bowl topsoil. The stories were political lessons, explaining differences in administrative agencies and the power of people to inspire resolutions. The stories were about greed replacing wilderness. The stories were about different visions of the past and future and how nature might fit in.

Conclusion: The Ends

ALTHOUGH MANY HISTORIANS WRITE narratives, history itself does not structure itself into neatly packaged stories, something this book affirms. To be sure, many traditional elements of a good story exist here. I found an outstanding setting, a cast of vivid and diverse characters, and rising tension. However, as much as I wanted a dramatic climax and clear resolution to the contest, such a denouement eluded this story. Kennecott pursued its plans; activists organized; government agencies watched. But never did a court stop the corporation from digging. No legislation nailed the casket lid shut over the top of the open pipe. No mass action convinced copper executives to cut their losses and leave the Suiattle River drainage. No president signed an executive order banning mining in designated wilderness

Instead the issue kindled, heated, and then long smoldered. Copper prices and production costs never seemed to match to favor Kennecott's bottom line. A sudden shift in either prices or expenses, though, meant an open-pit mine at Miners Ridge always remained a possibility. In 1977 the North Cascades Conservation Council newsletter reminded N3C members of Kennecott's continued presence and threat. Fortunately, no road slithered all the way up the Suiattle Valley, but helicopters kept bringing prospectors to Miners Ridge, leaving "the Cascades' largest garbage dump—miles of plastic pipe, sheets of plywood, rusting oil drums, hardened sacks of cement, wire, lumber, and old clothing scattered everywhere throughout the meadows of this fragile alpine country." Kennecott remained a "sleeping giant" that, if roused to wake, could finally scrape off Plummer Mountain's shoulder. Be prepared, the N3C warned, to combat the threat again.[1] But the giant did not wake. And because it never woke, the narrative—and its dramatic tension—mostly dropped off.

There is no mine on Miners Ridge and never will be. Kennecott Copper Corporation sold its claim in the 1986 to the Chelan County Public Utility

District (PUD), which used the site to gather snowfall data. Then the federal Wild Sky Wilderness bill that passed in 2008 allowed for a land swap between the PUD and the Forest Service. Because remnants from mining exploration activity might turn out to be harmful and require costly clean up, the Forest Service acquired only a conservation easement, not outright ownership. In 2010 the long-contested mining claims returned to Forest Service control, managed under Wilderness Act rules. When that happened, Gary Paull, the ranger who coordinated recreation for the Mt. Baker–Snoqualmie National Forest, told newspapers, "The Kennecott mine was one of my greatest fears as a kid, and [I] have been following it ever since—nudging and prodding from the inside hoping to see this outcome." He was in good and long-standing company. This was a defining campaign, not only in Paull's life but throughout the northwestern conservation community.[2]

I wish I could point definitively to why this episode resolved as it did. For the narrative, I would have preferred a special act of Congress to reform the Wilderness Act or exempt Glacier Peak Wilderness Area from the mining loophole. Or it would have been satisfying to see the Forest Service, its wound still fresh from having North Cascades National Park ripped from its control, institute such strict regulations that Kennecott just gave up. Better yet, I would have been gratified if the public testimony and actions of citizens had persuaded the corporation to redefine its notion of public good and reassess the merit of its North Cascades proposal. But no such turn of events happened; no such clarity explains the outcome. Yet this very elusiveness well represents the contingent nature of history, with all its incomplete evidence, memories, and conjectures—as well as a lack of moral certainty.

I have encountered some theories that others use to explain this episode. One of the most intriguing ideas is that a conspiring Kennecott Copper sought to capitalize on the height of the environmental era and get the federal government to buy its claim for top dollar. There is no real evidence for this in the record, although Representative Lloyd Meeds and others believed it to be true.[3] One source explained that Kennecott justified its abandonment because of strict regulations the Forest Service planned to impose, although agency evidence is circumstantial because it issued no regulations (since the company never formally applied for the permits it needed).[4] Another theory is that Scoop Jackson let Kennecott know that if it pushed ahead with this mine, it would be time for Congress to reconsider the 1872 Mining Act, the notoriously pro-mining law, any revision of which would hurt the mining industry in untold ways. Such a threat does not appear in writing.[5] Finally, those of an activist bent want to show how grassroots organizing by individuals and organizations stared down the mighty copper company and sent it scurrying out of the mountains.[6] The evidence for this is inconclusive, especially when framed as a head-to-head battle that activists won. A comparatively weak and unsteady economic argument for the mine seems as powerful a reason as any to explain Plummer Mountain remaining intact. So to present a tidy resolution attributed to a single factor would misread the past. And it would at least partly misread the purpose of this history and book.

The story I have told here details and explores an intense and inventive period of wilderness activism, peopled with a Supreme Court justice, a cabinet secretary, attorneys, doctors, and students. Letters to local papers and to Washington, D.C., conveyed northwesterners' concerns to Forest Service managers, Kennecott executives, and politicians. Journalists and authors told the story of this threatened mountainside. Hearings conducted by legislators gave citizens an occasion to participate in American democracy and revealed changing public sentiment. As much as anything, then, the story of Miners Ridge and Glacier Peak Wilderness Area and the North Cascades is a story of public citizenship, of engaging actively with a place and with people and with structures of power to find resolution and to stop what everyone outside the copper company believed was simple desecration.

In retrospect, Harvey Manning minimized the threat. The author of the Sierra Club book *The Wild Cascades,* writer of regional hiking guides, and longtime editor of the North Cascades Conservation Council newsletter concluded, "In the real world of the 1960s, there was clearly no chance that the mine ever would happen."[7] That assessment, about forty years after the fact, belied the real concern and profound uncertainty in 1966 when Kennecott announced its intent and when the law seemed clear, the Forest Service powerless, and Congress unwilling to stop the corporation. In hindsight, the mine might more clearly seem a longshot, as Manning suggested. However, reading public statements and private correspondence from the midst of the battle—including Manning's own—it is obvious that people believed Glacier Peak Wilderness Area was at risk. And that risk prompted fervent action.

Consider Brock Evans's experience and his own retrospective assessment. In the summer of 1965, the midpoint of a transformative decade, Evans and his wife Rachel traipsed and trudged through the forests, among the meadows, and across the ridges of Glacier Peak Wilderness Area for eight days, covering sixty miles. When they started, at the end of the road, they found a trail that led through "a primeval forest of huge trees." As they climbed thousands of feet, they emerged "in a sea of flowers on a level ridge." Their view swept "over an ocean of jagged peaks" and rested "on the immense white-rising bulk of Glacier Peak." Then, as now, streams tumble off the mountain and "wheel and arc around the great mountain to glacial headwaters on the other side." They could see it all, "a great circle and cycle of cloud, summit, meadow, river, flower, and forest . . . and the trail leads on, to a jewel lake, and from there to other places, just as beautiful, just as little known." The trip—a "profound experience"—awakened Evans to all that might be lost, for he and Rachel saw miners working, which filled them with "alarm and dread." They returned the following summer for what they expected to be "the very last time" and carried with them "very heavy and aching hearts," a burden made worse by helicopters carrying in people and mining equipment several times a day, a noisy interlude "accenting the feeling of dread" overcoming Evans as he picked his way through "the claims markers and road survey stakes."[8] Evans did not hide in his despair, did not disappear in his dread. Instead he became one of the central organizers among a growing troop of activists who worked to keep Kennecott away and protect wilderness.

Evans and his compatriots felt, in the middle of the 1960s, a palpable sense of near-inevitable doom. In a 2015 essay, almost fifty years after he started working for the Sierra Club as its Northwest representative, Evans wrote, "*It was a matter of life or death,* permanent loss or permanent rescue. We stood and fought, because whether or not a beautiful forest or river was protected was, in those desperate times, a matter of survival for that place." In explaining his generation's actions on behalf of wild nature, Evans moved toward metaphors of mortality. He also offered crucial context about conservationists' political weakness. When fighting Kennecott, activists like Evans and his friends only had the Wilderness Act. The National Environmental Policy Act, the Endangered Species Act, and the National Forest Management Act then existed only as activists' and legislators' dreams, not as democratic tools to solve problems.[9] And the Wilderness Act was brand new, and no one yet knew how far it might stretch before reaching its limits. So, with meager legal and political weapons, Evans—along with Polly Dyer and Grant McConnell, Mike McCloskey and David Brower, Fred Darvill and Ben Shaine, and so many others—raised a ruckus in public for wilderness.

Wilderness, in the time since Kennecott threatened an open pit in the North Cascades, has become a sometimes-fraught political cause and idea. Predictably, extractive commercial interests and their political supporters see wilderness, now as then, as an economic waste, with an irritating set of restrictions. But others also find much in wilderness to criticize. Some, noting the long human habitation on the continent, view wilderness as denying or belittling Indigenous presence on the land. Similarly, critics identify other environmental issues that require more attention, including those that disproportionately affect communities of color, making wilderness seem to be merely a passion of the privileged. Others argue that climate change has alerted us to the human impact that pervades the earth from the atmosphere to zooplankton; to speak of something "wild" in such a human-dominated world risks seeming naive. Still others see poor scientific rationale for wilderness protection as it developed in these early years of the Wilderness Act. There is merit in all these criticisms, worthy of the debates that reverberate in academic and activist communities.[10]

Yet my point here is not to interrogate wilderness as a philosophical idea or a political cause for today but to understand how Evans and his contemporary conservationists understood the wilderness issue at that time, from within the movement. As Evans told the story, his generation fought for wilderness because they loved it. They loved specific pieces of it that they knew, not some notional ideal, but specific places like Glacier Peak or Miners Ridge; and they hoped to save it from specific threats—timber sales, dams, mines. To garner protection, they knew they had to tell a story to convince the public and decision makers to preserve those places they loved. Evans pointed out that science was neither mature enough in 1967 nor persuasive enough for policymakers and the public to rally for such causes in the days before "biodiversity" and "ecosystem" resonated widely. He acknowledged that conservationists' narratives almost seem quaint in retrospect: "Odd as it may sound today, we emphasized how beautiful the forests . . . were, how rare and becoming rarer, and how population growth and recreation were overwhelming everything: Thus we needed to protect such places, now and forever."[11] Part of what Evans pointed out here was a reminder that wilderness was a political project and, as such, it required building a constituency. In 1967 that meant telling a story that sparked support, and in the world of the North Cascades, Kennecott became a useful antagonist that generated narrative conflict.

These narrative tensions reflected broader functions that wilderness serves. In his definitive history of environmental politics since the Wilderness Act, historian James Morton Turner concluded, "Wilderness means more than pristine wild lands, backpacking adventures, or a stronghold for biodiversity; wilderness also means engaging citizens—both for and against wild lands protections—in a sustained discussion toward the common interest. All of that is the promise of wilderness."[12] The campaign to stop Kennecott illustrate Turner's point perfectly. Like a magnet to iron filings, Miners Ridge attracted citizens who engaged—on the trail, in boardrooms, at hearings, in letters to the editor, through private correspondence. And appeals to the public good proved remarkably common, far more frequent than claims about untouched land or wildlife preservation or recreation opportunities.

Certainly Miners Ridge possesses historical resonance and importance beyond just this. In a big historical picture, it may be most significant because it became, according to natural resource law specialist John D. Leshy, "a watershed in implementation of the Wilderness Act's compromise on mining in wilderness." Never again did a mining corporation seriously attempt to use the loophole.[13] The efforts on behalf of Glacier Peak Wilderness Area demonstrated the power the public wielded and its overwhelming preference regarding the law's loophole. The cost of doing business in wilderness for mining corporations increased. Miners Ridge also played a supporting role in the drama associated with creating North Cascades National Park. Although writers have acknowledged it, they have done so only in passing.[14] The campaign against Kennecott folded into the efforts to make the North Cascades a national park. But because the final park boundaries excluded Miners Ridge and Glacier Peak, their role in building enthusiasm has been forgotten.

Of course, the history of Miners Ridge is most important to the place itself. Although it is important to see singular places' connections to broad national or global trends, such places deserve to have their histories told. And environmental activism, especially in this era, often centered on such specificity. What pulled citizens into a cause were concrete places that mattered to them. Again, Brock Evans is instructive. On July 18, 1967, he wrote to Kennecott's president, Frank Milliken. Evans was courteous and sought common ground. Like Milliken, Evans acknowledged the nation's mineral needs and searched for a balance in using land and other resources. But he drew a distinction. Wilderness and scenic beauty also were scarce and needed careful stewardship and protection. That is what Evans tried to explain to Milliken: "[W]e do not believe that we have yet reached the point where *in this place,* the scenic and wilderness value is transcended by the mineral value. In other words, our view is that proper utilization of *this place* is for the other two of its three primary natural resources—the scenic beauty and the wilderness."[15] *In this place*—Miners Ridge—Evans pointed out that wilderness mattered more than minerals.

And *for* this place, this beautiful and wild mountain ridge, people worked creatively and energetically, with dedication and urgency. High-ranking officials, professional conservationists, and grassroots activists climbed together to secure the wilderness summit. To keep Miners Ridge whole meant clarifying conservation's purpose. In the late 1960s, for conservationists based in the Northwest and across the country, conservation's meaning included wilderness, a place protected from commercial extraction, despite what Congress had negotiated into the Wilderness Act. In the process, those involved debated and *created* the public interest, for it is in engagement as citizens that the public interest is defined. This was—and is—the means and the ends to wilderness.

NOTES

INTRODUCTION

1. A copy of the advertisement is available in Carton 46, Folder 9, GMP.
2. McPhee, *Encounters with the Archdruid*, 36.
3. Michael McCloskey, *In the Thick of It*, 94; McPhee, *Encounters with the Archdruid*, 37–38; Wyss, *Man Who Built the Sierra Club*, 208–14. Brower did not create the ad described above ("An Open Pit"), and he resented McPhee for implying that he had and the suggestion that he or the Sierra Club exaggerated. Turner, *David Brower*, 158–59.
4. McPhee, *Encounters with the Archdruid*, 37, 38.
5. Joel Connelly, "The Open Pit Is Finally Put Away," *Seattle Post-Intelligencer*, May 5, 2010, https://blog.seattlepi.com/seattlepolitics/2010/05/05/the-open-pit-is-finally-put-away.
6. Consider the short introduction to the issue on the North Cascades Conservation Council website as an example: North Cascades Conservation Council, "Kennecott Open Pit on Miner's Ridge: A Lasting Victory," in "Mining Issues," http://www.northcascades.org/wordpress/mining. A local newspaper's summary when the issue finally resolved: Bill Sheets, "Swap Adds Miner's Ridge to Glacier Peak Wilderness," *HeraldNet.com* (Everett, Wash.), May 6, 2010, https://www.heraldnet.com/news/swap-adds-miners-ridge-to-glacier-peak-wilderness.
7. Cronon, "Uses of Environmental History," 13.
8. People debate the boundaries of the North Cascades. The most expansive definition includes all of Washington's Cascades. Others place the rough boundary at Mount Rainier or Snoqualmie Pass. Using Stevens Pass as the southern boundary makes for a smaller region, but it provides a satisfying historical unity, demarking the area most commonly and consistently referred to as the North Cascades.
9. McConnell, "The Cascade Range," offers a sweeping overview of the range. For geology, begin with Tabor and Haugerud, *Geology of the North Cascades*.
10. The proper name is Miners Ridge, but many referred to it in writing as Miner's Ridge, with the apostrophe.

11. "Scenic climax" became a common term tossed around by conservationists during this campaign. I cannot locate its original use.
12. Darvill, *Hiking the North Cascades*, 263–66; Spring and Manning, *100 Hikes in Washington's North Cascades*, 50–53; Tabor and Haugerud, *Geology of the North Cascades*, 106–7.
13. A persuasive argument for the material foundations of the past is found in LeCain, *Matter of History*.
14. Nicolson, *Mountain Gloom and Mountain Glory*.
15. The literature on wilderness history is vast, but two classics to start with are Cronon, "The Trouble with Wilderness" and Nash, *Wilderness and the American Mind*. Recent scholarship has deepened and challenged these approaches, but their broad outlines still work. A useful history of camping is Young, *Heading Out*.
16. Marsh, *Drawing Lines in the Forest*, esp. 1–18.
17. I am adapting Keith Basso's provocative title on a suggestion from Jeff Sanders. Basso, *Wisdom Sits in Places*.
18. Sometimes literally—see, for example, Marsh, *Drawing Lines in the Forest*, 163n50.
19. Harvey, *Wilderness Forever*, 187, 202–44.
20. Harvey, *Wilderness Forever*, 204, 237 on the mining compromise; Schulte, *Wayne Aspinall and the Shaping of the American West*, 115–62, on Aspinall's role in passing the bill. Scott, *Enduring Wilderness*, 47–55 on the legislative process and compromises. See also Wilderness Act, Public Law 88-577, *Statutes at Large*, 78 (1964): 890–97.
21. Although they do not use the term "path dependency," both Charles F. Wilkinson and Christopher McGrory Klyza demonstrate the power of early policy decisions in framing public land possibilities long after initial enactments. Wilkinson, *Crossing the Next Meridian*; Klyza, *Who Controls Public Lands?*
22. Although great fear about the mining exception animated conservationists, it does not seem to have been invoked successfully. Leshy, *The Mining Law*, 235; Scott, *Enduring Wilderness*, Kindle edition, location 2358.
23. Cronon, "Uses of Environmental History," 14–16.
24. An excellent study of the influence of unbuilt developments is Peyton, *Unbuilt Environments*.

CHAPTER 1. THE FOUNDATION

1. See, for instance, Williams, *U.S. Forest Service in the Pacific Northwest*, 145; Caldbick, "Frederick William Cleator."
2. Cleator, "Report on Glacier Peak Wilderness Area," in Box: Division of Recreation and Lands—N. Cascades (095-65B215 1–2, Box 6/25871), Folder: Glacier Peak Wilderness Area Wenatchee, 6/13/36–12/3/48), RUSFS, 1.
3. Sutter, *Driven Wild*, see roads and automobiles as the unifying factor for the interwar origins of the wilderness movement, and Rogers, *Roads in the Wilderness*, focuses on roads as the site of conflict in Utah throughout the twentieth century.

4. Danner, *Crown Jewel Wilderness*, 28–34, captures the story of this mostly forgotten "Ice Peaks National Park." See also Louter, *Windshield Wilderness*, 112.
5. Jagodinsky, *Legal Codes and Talking Trees*, 217.
6. A quick examination of any map of the area shows the river drainages as described. But see also overviews such as Tabor and Haugerud, *Geology of the North Cascades*, for introductory geology, and for basic history and geography see McConnell, "Cascade Range," esp. 87–96.
7. Drawn from Jagondinsky, *Legal Codes and Talking Trees*, 216–20. See also Marsh, "Ups and Downs of Mountain Life," 201–3.
8. Fiege, *Republic of Nature*, 156–98.
9. Jagondinsky, *Legal Codes and Talking Trees*, 220–22.
10. Hurst, *Law and the Condition of Freedom;* White, "Contested Terrain," esp. 197–99.
11. Marsh, "Ups and Downs of Mountain Life," 207; Wyckoff and Dilsaver, "Defining the Mountainous West."
12. The law is "An Act to Provide the Development of the Mining Resources of the United States," *US Statutes at Large* 17 (May 10, 1872), 91–96, quotation from 91 with emphasis added. See also Brosnan, "Law and the Environment," 528–30; Gates, *History of Public Land Law Development*, 723; Klyza, *Who Controls Public Lands?* 27–47; Wilkinson, *Crossing the Next Meridian*, 28–74.
13. Beckey, *Range of Glaciers*, 323–61; Danner, *Crown Jewel Wilderness*, 13–14; Dietrich, *North Cascades*, 88–91.
14. Woodhouse, *Monte Cristo*.
15. Information about the Holden Mine comes from various sources, including Adams, *Holden Mine;* Beckey, *Range of Glaciers*, 330; Edmund, *Highlights of the Holden Copper Mine;* Manning and North Cascades Conservation Council, *Wilderness Alps*, 43–44.
16. Beckey, *Range of Glaciers*, 329.
17. Standard accounts of Yellowstone's creation are Nash, *Wilderness and the American Mind*, 108–21; Runte, *National Parks*, 29–41. The quotation is from "An Act to Set Apart a Certain Tract of Land Lying Near the Headwaters of the Yellowstone River as a Public Park," in Dilsaver, *America's National Park System*.
18. Steen's general history of the Forest Service explains this background. See Steen, *U.S. Forest Service*.
19. The Washington Forest Reserve history in the context of Cleveland's actions is constructed from Williams, *U.S. Forest Service*, 39–43.
20. Quoted in Williams, *U.S. Forest Service*, 54. Besides the natural and transportation obstacles in the public land, private forests met most of the region's lumber requirements.
21. The 1907 law that changed the name from forest reserves to national forests also stripped the president of the power to create the forests. Henceforth only Congress could create national forests. President Roosevelt, just before the law went into effect, reserved some sixteen million acres in new national forests. See Williams, *U.S. Forest Service in the Pacific Northwest*, 62.
22. Quoted in Williams, *U.S. Forest Service in the Pacific Northwest*, 60.

23. The "greater good" phrase appears in many places, including Pinchot, *Breaking New Ground*, 462.
24. Nash, *Wilderness and the American Mind*, and Runte, *National Parks*, are the classic accounts. See also Cronon, "Trouble with Wilderness."
25. Chamberlin, *On the Trail;* Kjeldsen, *Mountaineers*, 11–13.
26. Wolfe, *John of the Mountains*, 317.
27. "An Act to Establish National Park Service and for Other Purposes," in Dilsaver, *America's National Park System.*
28. Rothman, "Regular Ding-Dong Fight."
29. For promotion of parks for railroads and automobiles, see Shaffer, *See America First;* for the circuit, see Rothman, *America's National Monuments;* for the pioneering use of automobiles in Mount Rainier, see Louter, *Windshield Wilderness*, 11–67.
30. On roads and automobiles, see Sutter, *Driven Wild;* for the ways a camping craze brought modernity into the woods, see Young, *Heading Out.*
31. Sutter, *Driven Wild*, 64–65.
32. Leopold, "Wilderness and Its Place in Forest Recreational Policy," 79. See also Sutter, *Driven Wild*, esp. 67–73.
33. Quoted in Sutter, *Driven Wild*, 72.
34. Leopold, "Wilderness and Its Place," 79.
35. The best account of Marshall's beliefs and his role in founding the Wilderness Society is Sutter, *Driven Wild*, esp. 194–238.
36. For more on Marshall's time in the Forest Service generally, see Sutter, *Driven Wild*, 221–27; Glover, *Wilderness Original.* For his work on the Wind River station, see Brock, *Money Trees*, 66–99.
37. Sutter, *Driven Wild*, 221–26; Glover, *Wilderness Original*, 144–48; *National Plan for American Forestry*, esp. 463–88.
38. Summarized in Sutter, *Driven Wild*, 253.
39. Louter, *Windshield Wilderness.*
40. Sutter, *Driven Wild*, 3–7.
41. [Marshall], "Three Great Western Wildernesses," 10. The article is unsigned, but excerpts from a letter from Marshall explained that he wrote it. "Some Early Glacier Peak Wilderness Area History," undated memo (ca. 1956) in Carton 46, Folder 5, GMP.
42. Danner, *Crown Jewel Wilderness*, 28–30. Watkins, *Righteous Pilgrim*, 562–68, covers Ickes and the Olympic Mountains, giving a taste of his maneuvering in the Pacific Northwest.
43. Danner, *Crown Jewel Wilderness*, 25–26.
44. [Marshall], "Three Great Western Wildernesses," 10; Danner, *Crown Jewel Wilderness*, 28.
45. Cleator, "Report on Glacier Peak Wilderness Area," 1.
46. Sutter, *Driven Wild*; Maher, *Nature's New Deal.*
47. Cleator, "Report on Glacier Peak Wilderness Area," 5.
48. See Marsh, *Drawing Lines in the Forest*, 131. He discusses this dynamic in a later period, but the insight is relevant to earlier decades.

49. Cleator, "Report on Glacier Peak Wilderness Area," 1–7, quotation from 1.
50. Cleator, "Report on Glacier Peak Wilderness Area," 8–8a.
51. Cleator, "Report on Glacier Peak Wilderness Area," 11.
52. Steen, *U.S. Forest Service*, 162; Williams, *U.S. Forest Service*, 145.
53. Leon F. Kneipp, Washington, D.C., to Regional Forester [Clarence J.] Buck, Portland, Oregon, September 9, 1936, Box: Division of Recreation and Lands—N. Cascades (095-65B215 1-2, Box 6/25871), Folder: Glacier Peak Wilderness Area Wenatchee, 6/13/36–12/3/48), RUSFS.
54. Kneipp to Buck.
55. F. A. Silcox, Washington, D.C., to [Clarence J.] Buck, Portland, Oregon, March 22, 1939, Box: Division of Recreation and Lands—N. Cascades (095-65B215 1–2, Box 6/25871), Folder: Glacier Peak Wilderness Area Wenatchee, 6/13/36—12/3/48), RUSFS. This letter references the rejection of the wilderness area in 1936; the memo it references is unavailable.
56. Glover, *Wilderness Original*, 56.
57. Cleator recalls his hike with Marshall in a document dated June 1957 graciously provided to me by Lauren Danner. The original is found in Box 28, Folder 3, WSR. Marshall's letter is quoted in Cleator's recollection.
58. Quoted in Glover, *Wilderness Original*, 264 (see also 248).
59. Glover, *Wilderness Original*, 248.
60. Robert Sterling Yard, Washington, D.C., to F. A. Silcox, Washington, D.C., July 22, 1938, Box: Division of Recreation and Lands—N. Cascades (095-65B215 1–2, Box 6/25871), Folder: Glacier Peak Wilderness Area Wenatchee, 6/13/36–12/3/48), RUSFS.
61. The quotation comes from Robert Sterling Yard, Washington, D.C., to F. A. Silcox, Washington, D.C., February 23, 1939, Box: Division of Recreation and Lands—N. Cascades (095-65B215 1–2, Box 6/25871), Folder: Glacier Peak Wilderness Area Wenatchee, 6/13/36–12/3/48), RUSFS. The correspondence includes several letters in these files.
62. F. A. Silcox, Washington, D.C., to Buck, March 22, 1939, Box: Division of Recreation and Lands—N. Cascades (095-65B215 1–2, Box 6/25871), Folder: Glacier Peak Wilderness Area Wenatchee, 6/13/36–12/3/48), RUSFS.
63. Danner, *Crown Jewel Wilderness*, 26; USDA, Forest Service, *Glacier Peak Land Management Study*, 1.
64. See, for example, F. V. Horton, Portland, Oregon, to Lyle Watts, August 2, 1940, Box: Division of Recreation and Lands—N. Cascades (095-65B215 1–2, Box 6/25871), Folder: Glacier Peak Wilderness Area Wenatchee, 6/13/36–12/3/48), RUSFS.
65. Manning and North Cascades Conservation Council, *Wilderness Alps*, 77.
66. Irving M. Clark, Bellevue, Washington, to John M. Coffee, Washington, D.C., May 11, 1940, Box: Division of Recreation and Lands—N. Cascades (095-65B215 1–2, Box 6/25871), Folder: Glacier Peak Wilderness Area Wenatchee, 6/13/36–12/3/48), RUSFS.
67. F. V. Horton, Memorandum for D, Box: Division of Recreation and Lands—N. Cascades (095-65B215 1–2, Box 6/25871), Folder: Glacier Peak Wilderness

Area Wenatchee, 6/13/36–12/3/48), RUSFS. For more on Clark, see Glover, *Wilderness Original*, 180, 189–91, 208–9; Kjeldsen, *Mountaineers*, 9, 120, 137; Manning and North Cascades Conservation Council, *Wilderness Alps*, 73–75. The Sierra Club followed a similar trajectory, in which politics suffused early activities but gave way to primarily outings before the club returned to political action. See Cohen, *History of the Sierra Club*.

68. Horton to Watts; John Sieker, Washington, D.C., to Irving M. Clark, Bellevue, Washington, May 17, 1940, and Irving M. Clark, Bellevue, Washington, to F. V. Horton, Portland, Oregon, August 23, 1940, both in Box: Division of Recreation and Lands—N. Cascades (095-65B215 1–2, Box 6/25871), Folder: Glacier Peak Wilderness Area Wenatchee, 6/13/36–12/3/48), RUSFS. For the context of Horton's local consultations, see Danner, *Crown Jewel Wilderness*, 31.
69. Horton to Watts.
70. This is not to say that there were no concerns about logging. Wilderness in the Cascades was tied up with wilderness, parks, and logging on the nearby Olympic Peninsula. However, because logging was obviously so central a question in later wilderness campaigns and because mining ultimately did *not* develop into an economic or political powerhouse in the region, it is important to remember that other possibilities existed.
71. C. M. Granger, Washington, D.C., to Mr. Yard, August 15, 1940, in Carton 229, Folder 30, SCMP-EWP.
72. On a copy of this letter in Forest Service files someone penciled "pretty thick" in the margins next to this. See Robert Sterling Yard, Washington, D.C., to C. M. Granger, Washington, D.C., October 15, 1940, Box: Division of Recreation and Lands—N. Cascades (095-65B215 1–2, Box 6/25871), Folder: Glacier Peak Wilderness Area Wenatchee, 6/13/36–12/3/48), RUSFS.
73. Yard to Granger.
74. Lyle F. Watts to Chief [C. M. Granger], Washington, D.C., October 30, 1940, and C. M. Granger, Washington, D.C., to Robert Sterling Yard, Washington, D.C., November 12, 1940, both in Box: Division of Recreation and Lands—N. Cascades (095-65B215 1–2, Box 6/25871), Folder: Glacier Peak Wilderness Area Wenatchee, 6/13/36–12/3/48), RUSFS.
75. Watts to Chief [Granger].
76. Horton to Watts refers to specific members of local chambers of commerce calling for this very thing.
77. See discussion, for example, in Sowards, *Environmental Justice*, 62–64. See also Schrepfer, *Nature's Altars*.
78. For two treatments of gear, see Turner, "From Woodcraft to 'Leave No Trace,'" and Young, *Camping Out*.
79. This is a major theme in Marsh, *Drawing Lines*.
80. Granger to Yard.
81. P. E. Oscarson, Spokane, Washington, to Regional Forester, Portland, Oregon, September 16, 1942, and Lyle F. Watts to Chief, Washington D.C., September 19, 1942, quotation, both in Box: Division of Recreation and Lands—N.

Cascades (095-65B215 1–2, Box 6/25871), Folder: Glacier Peak Wilderness Area Wenatchee, 6/13/36–12/3/48), RUSFS.

82. Steen, *U.S. Forest Service*, 137–40; Lewis, *Forest Service and the Greatest Good*, 246.

83. Early [*sic*] H. Clapp, Washington, D.C., to Director, Bureau of Mines, September 23, 1942, Box: Division of Recreation and Lands—N. Cascades (095-65B215 1–2, Box 6/25871), Folder: Glacier Peak Wilderness Area Wenatchee, 6/13/36–12/3/48), RUSFS.

84. Clapp to Director, Bureau of Mines. Clapp's point here about "irreparable impairment" contradicts later points made that the mine would be minimally intrusive.

85. Frank A. Ayer to C. M. Granger, Washington, D.C., September 24, 1942, Box: Division of Recreation and Lands—N. Cascades (095-65B215 1–2, Box 6/25871), Folder: Glacier Peak Wilderness Area Wenatchee, 6/13/36–12/3/48), RUSFS.

86. C. M. Granger, Washington, D.C., to Frank A. Ayer, Washington, D.C., September 25, 1942; Kneipp, Washington, D.C., to Forest Service, Portland, OR, September 25, telegram; and L. F. Kneipp, Washington, D.C., to Irving M. Clark, Bellevue, Washington, November 7, 1942: all in Box: Division of Recreation and Lands—N. Cascades (095-65B215 1–2, Box 6/25871), Folder: Glacier Peak Wilderness Area Wenatchee, 6/13/36–12/3/48), RUSFS.

87. Kneipp to Forest Service, telegram; Lyle F. Watts, to Irving M. Clark, Bellevue, Washington, October 2, 1942, Box: Division of Recreation and Lands—N. Cascades (095-65B215 1–2, Box 6/25871), Folder: Glacier Peak Wilderness Area Wenatchee, 6/13/36–12/3/48), RUSFS.

88. C. M. Granger, Washington, D.C., to Robert Sterling Yard, Washington, D.C., October 9, 1942, Box: Division of Recreation and Lands—N. Cascades (095-65B215 1–2, Box 6/25871), Folder: Glacier Peak Wilderness Area Wenatchee, 6/13/36–12/3/48), RUSFS.

89. Watts to Clark, October 2, 1942.

90. Yard, Washington, D.C., to C. M. Granger, Washington, D.C., October 22, 1942, Box: Division of Recreation and Lands—N. Cascades (095-65B215 1–2, Box 6/25871), Folder: Glacier Peak Wilderness Area Wenatchee, 6/13/36–12/3/48), RUSFS.

91. Irving M. Clark, Bellevue, Washington, to Chief, U.S. Forest Service, Washington, D.C., October 24, 1942; Irving M. Clark, Bellevue, Washington, to Regional Forester, Portland Oregon, June 24, 1943; and Irving M. Clark, Bellevue, Washington, to Regional Forester, Portland, Oregon, July 3, 1945: all in Box: Division of Recreation and Lands—N. Cascades (095-65B215 1–2, Box 6/25871), Folder: Glacier Peak Wilderness Area Wenatchee, 6/13/36–12/3/48), RUSFS.

92. H. J. Andrews, to Irving M. Clark, Bellevue, Washington, July 12, 1945, Box: Division of Recreation and Lands—N. Cascades (095-65B215 1–2, Box 6/25871), Folder: Glacier Peak Wilderness Area Wenatchee, 6/13/36–12/3/48), RUSFS.

93. Andrews to Clark.

CHAPTER 2. THE WILDERNESS

1. Background on McConnell can be found in McConnell, "Conservation and Politics in the North Cascades," quotation from 4. See also McConnell, *Stehekin*, esp. 207–11 (afterword written by McConnell's granddaughter); Manning and North Cascades Conservation Council, *Wilderness Alps*, 85–86; and Marsh, *Drawing Lines*, 41–42, 49–51.
2. McConnell, "Cascades Wilderness," 24, 28, 29. The superlative nature of the landscape was often noted using the very term "superlative." See, for instance, Harold Bradley, "Northern Cascades," 28.
3. McConnell, "Cascades Wilderness," 26, 29, 30, 31. Wood explored the emergence of holistic ideas in the postwar era in *More Perfect Union*.
4. McConnell, "Conservation Movement," 477.
5. McConnell, "Cascades Wilderness," 31. For more on zoning these lands, see Marsh, *Drawing Lines*, 57.
6. Hays, *Beauty, Health, and Permanence*, offers a classic account of this transformation.
7. Hirt, *Conspiracy of Optimism*, explains this shift.
8. For defense to offense, see Michael McCloskey, "Wilderness Movement at the Crossroads"; and Nash, *Wilderness and the American Mind*, 222.
9. Marsh covers the Three Sisters campaign expertly in *Drawing Lines*, 19–37. For the best overview on the wilderness bill campaign, see Harvey, *Wilderness Forever* (152–69 on reclassification). See also Scott, *Enduring Wilderness*, 35–36.
10. Henry A. Harrison to William Gissberg, Marysville, Washington, November 29, 1954, Box: 95-65B215 1-2 Box 5-25870, Folder: Near-Natural Area, Glacier Peak Wilderness Area Wenatchee, 1950 to 1955, RUSFS. Emphasis added.
11. H. J. Engles, "Report on that Portion of the Glacier Peak Wilderness Area Located in the Darrington Ranger District," October 12, 1950; and Recreation and Lands, from Supervisor, Mt. Baker, H. Phil Brandner, Bellingham, WA, October 30, 1950: both in Box: 95-65B215 1-2 Box 5-25870, Folder: Near-Natural Area, Glacier Peak Wilderness Area Wenatchee, 1950 to 1955, RUSFS.
12. "Report for Wilderness Area," October 30, 1950, Bellingham, Washington; "Report on that Part of the Glacier Peak Wilderness Area within the Suiattle Range District," October 13, 1950, Darrington, Washington; and H. Phil Brandner, to Margaret B. Callahan, Seattle, April 15, 1953: all in Box: 95-65B215 1-2 Box 5-25870, Folder: Near-Natural Area, Glacier Peak Wilderness Area Wenatchee, 1950 to 1955, RUSFS.
13. Harrison to Gissberg.
14. "Report on Proposed Glacier Peak and Snow King Wilderness Area," Mt. Baker National Forest, June 14, 1956; Box: 95-65B215 1-2 Box 5-25870), Folder: Near-Natural Area, Glacier Peak Wilderness Area Wenatchee National Forest, 1956 to 1957, RUSFS.
15. "Report on Proposed Glacier Peak."
16. See, for example, Manning and North Cascades Conservation Council, *Wilderness Alps*, 79–121; Marsh, *Drawing Lines*.

17. "Report on Proposed Glacier Peak," 6–9.
18. "Report on Proposed Glacier Peak," 5–6, 11, quotations from 6.
19. USDA, Forest Service, *Glacier Peak Land Management Study*, 1–2.
20. "Report on Proposed Glacier Peak," 1.
21. USDA, Forest Service, *Glacier Peak Land Management Study*, 2–3, 6, 4, 9.
22. J. Herbert Stone's public notice appeared appended to the proposal. USDA, Forest Service, *Glacier Peak Wilderness Proposal.*
23. *Glacier Peak Proposal*, 1–3.
24. *Glacier Peak Proposal*, 2; [Stone], "Glacier Peak Wilderness Proposal," 23. For broader context, see Marsh, *Drawing Lines*, 49–54; Manning and North Cascades Conservation Council, *Wilderness Alps*, 105–21; Danner, *Crown Jewel Wilderness*, 79–83.
25. [Stone], "Glacier Peak Wilderness Proposal."
26. See Chamberlin, *On the Trail*; Schrepfer, *Nature's Altars*.
27. Grauer, "Mazamas."
28. Kjeldsen, *Mountaineers*, esp. 11–19, 115–38, quotation from 118; Manning and North Cascades Conservation Council, *Wilderness Alps*, 53, 88–90; Arntz, *Extraordinary Women Conservationists of Washington*, 29–35, 37–44.
29. Manning and North Cascades Conservation Council, *Wilderness Alps*, 93.
30. Philip H. Zalesky, "Glacier Peak Area: Wilderness or Waste?," *Mountaineer* 48, no. 13 (December 1955): 37–40. On Marshall in this vein, see Sutter, *Driven Wild*, 194–238.
31. Danner, *Crown Jewel Wilderness*, 44–45; Manning and North Cascades Conservation Council, *Wilderness Alps*, 92–97.
32. Chamberlin, *On the Trail*, emphasizes the importance of communalism in developing American hiking culture.
33. See Cohen, *History of the Sierra Club*, 19–20 (for the origin of the high trips), 89–100 (for the questioning of their ethics).
34. Cohen, *History of Sierra Club*, 221.
35. Rosters of simultaneous trips to Glacier Peak, for various skill levels and with various time commitments, reveal that more Californians than northwesterners participated. Rosters are found in Carton 72, Folder 18, Sierra Club Records, Bancroft Library, University of California, Berkeley. Zahnisers and Browers: Harvey, *Wilderness Forever*, 192.
36. Biographies of Zahniser and Brower reveal their often hectic travel schedules. See Harvey, *Wilderness Forever*; Turner, *David Brower*; Wyss, *Man Who Built the Sierra Club.*
37. "Our Greatest Wilderness Park-Land," unpaginated photographs. One camper was Charles A. Reich, a young lawyer just off a term clerking for Supreme Court justice Hugo Black. At the end of the next decade, he published a counterculture landmark, *The Greening of America.*
38. Warth, "Glacier Peak Wilderness," 173, 174, 175.
39. Warth, "Glacier Peak Wilderness," 193.
40. Danner, *Crown Jewel Wilderness*, 55–62; Manning and North Cascades Conservation Council, *Wilderness Alps*, 109–11.

41. Manning reported that this was an unprecedented number. Without citation, this claim is hard to verify. Manning and North Cascades Conservation Council, *Wilderness Alps*, 121.
42. E. L. Peterson, "Decision of the Secretary of Agriculture Establishing the Glacier Peak Wilderness Area Mt. Baker and Wenatchee National Forests, Washington," September 6, 1960, in Folder: Kennecott Mine—Speeches and Writings, misc. 1960–1970, Box 8, N3CR. Notwithstanding the outcome, several speakers at the Wenatchee hearing who worked for Kennecott subsidiaries, and one person who refused to answer whether he did, spoke out against the proposed wilderness. They feared restrictive regulations, and one viewed the wilderness advocates as a special interest: "We are in a strange place where the mining and timber companies are bending over backwards and the self-styled conservationists are in the position of being the land-grabbers for their special purposes." Excerpts from the hearings containing geologists' take on wilderness are found in "About Kennecott: How Its Hirelings Spoke against the Glacier Peak Wilderness," 16.
43. Decision of the Secretary of Agriculture Establishing the Glacier Peak Wilderness Area, Mt. Baker and Wenatchee National Forests, Washington [draft]; Box: 95-65B215 1-2. Box 5-25870; Folder: Near-Natural Area, Glacier Peak Wilderness Area Wenatchee National Forest, 1960, RUSFS.
44. Richard E. McArdle, Washington D.C., to Regional Forester, Region 6, July 29, 1960, Box: 95-65B215 1–2, Box 5-25870, Folder: Near-Natural Area, Glacier Peak Wilderness Area Wenatchee National Forest, 1960, RUSFS.
45. Douglas, "America's Vanishing Wilderness," 37–41, 77, quotation from 77. For more on his concept in action, see Sowards, *Environmental Justice*, 2–3, 81–111.
46. Sierra Club, *Wilderness Alps of Stehekin*; Wyss, *Man Who Built the Sierra Club*, 100–101; Turner, *David Brower*, 116.
47. Manning and North Cascades Conservation Council, *Wilderness Alps*, 108.
48. Harvey, *Wilderness Forever*, 182.
49. Turner, *David Brower*, 38–39, 54–55, 93–102; "be bottled" is Brower's son Ken's phrase, quoted on 94; Wyss, *Man Who Built the Sierra Club*, 52, 153–54.
50. Turner, *David Brower*, 96–97; Wyss, *Man Who Built the Sierra Club*, 153. I used an inflation calculator to convert prices: https://www.calcxml.com/do/bud12 (accessed February 28, 2018).
51. Harvey Manning, Issaquah, Washington, to George Marshall, Los Angeles, May 5, 1965, Carton 5, Folder 27, GMP.
52. Manning to Marshall, May 5, 1965; Manning, *Wild Cascades*, 11 (for the production schedule); Manning and North Cascades Conservation Council, *Wilderness Alps*, 129; Danner, *Crown Jewel Wilderness*, 144–45.
53. Manning, *Wild Cascades*.
54. Sowards, *Environmental Justice*, 16–18, on how Douglas used frontier tropes; an excerpt of Douglas's review is blurbed on the back cover of the fifth edition of Nash, *Wilderness and the American Mind*.
55. Douglas, foreword to Manning, *Wild Cascades*, 15. Douglas had some misgivings about the national park, worrying that park development might wreck the wilderness. See, for instance, his letter to the NPS director in 1967: William O.

Douglas to George B. Hartzog Jr., Washington, D.C., March 9, 1967, Box 558, Folder: North Cascade Mountains (1967–1968), WODP.

56. Manning, *Wild Cascades*, 17.
57. Manning, *Wild Cascades*, 69–70.
58. Manning, *Wild Cascades*, 103 (first quotation), 104 (second quotation), 89–90 (wildlife).
59. Manning, *Wild Cascades*, 107, 108.
60. Manning, *Wild Cascades*, 109, 110.
61. Manning, *Wild Cascades*, 114–17; Danner, *Crown Jewel Wilderness*, 26.
62. Manning, *Wild Cascades*, 117.
63. Cohen, *History of the Sierra Club*, 292; Danner, *Crown Jewel Wilderness*, 144–45.
64. Quoted in Manning and North Cascades Conservation Council, *Wilderness Alps*, 129. The newspaper and its publisher Rufus Woods were leading advocates for Grand Coulee Dam. See Ficken, *Rufus Woods*.
65. George Marshall, Los Angeles, to Harvey Manning, Issaquah, Washington, August 30, 1965, Carton 5, Folder 27, GMP.

CHAPTER 3. THE DANCE

1. Chriswell, *Memoirs*; Williams, *U.S. Forest Service*, 148, 153, 181–82; Manning and North Cascades Conservation Council, *Wilderness Alps*, 124. For "getting out the cut" as the institutional context, see Hirt, *Conspiracy of Optimism*, 131–50.
2. Walt Woodward, "Open-Pit Copper Mine Possible in Proposed Park," *Seattle Times*, December 22, 1966.
3. Wilderness Act, Public Law 88-577, *US Statutes at Large* 78 (1964), 894.
4. William A. Worf, Washington, D.C., to R. J. Costley, Washington, D.C., January 25, 1966, Box: Region 6, Division of Recreation and Lands Recreation Studies (095-76B2307 Box 3-1990), Folder: Wildernesses and Primitive Areas Glacier Peak Wilderness, Folder #3B Mineral Prospecting and Mining, RUSFS.
5. Mrs. William Devin, Bremerton, Washington, to Stuart [*sic*] Udall, Washington, D.C., August 9, 1966, Box: Region 6, Division of Recreation and Lands Recreation Studies (095-76B2307 Box 3-1990); Folder: Wildernesses and Primitive Areas Glacier Peak Wilderness, Folder 3a, RUSFS.
6. Forest Supervisor, Mt. Baker, to Regional Forester, August 30, 1966, Box: Region 6, Division of Recreation and Lands Recreation Studies (095-76B2307 Box 3-1990), Folder: Wildernesses and Primitive Areas Glacier Peak Wilderness, Folder 3B (Mineral Prospecting and Mining), RUSFS.
7. A series of communiques shows how the agency struggled with its own regulations, how they fit with the new Wilderness Act, and how to apply them to Kennecott's subsidiary. Besides the sources cited in the preceding note, see also "Summary of Pertinent Data regarding Mining Claims of Kennecott Copper Corp. in the Miners Ridge Area, Glacier Peak," September 29, 1966, Forest Supervisor, Mt. Baker, to Regional Forester, September, 1966; Philip L. Heaton, to Regional Attorney, September 30, 1966; C. C. Carlson, Portland, Oregon, to Philip L. Heaton, October 6, 1966; R. H. Payne, Washington, D.C., to Regional Forester, R-6, October 14, 1966: all in Box: Region 6, Division of Recreation and

Lands Recreation Studies (095-76B2307 Box 3-1990), Folder: Wildernesses and Primitive Areas Glacier Peak Wilderness, Folder #3B Mineral Prospecting and Mining, RUSFS.

8. Patrick D. Goldsworthy, Seattle, to J. Herbert Stone, Portland, Oregon, July 26, 1966, Box 8, Folder: Kennecott Mine General Correspondence 1966, N3CR.
9. Goldsworthy to Stone, July 26, 1966.
10. Henry M. Jackson, Washington, D.C., to Patrick D. Goldsworthy, Seattle, August 1, 1966; Edward C. Crafts, Washington, D.C., to Patrick D. Goldsworthy, August 4, 1966; A. E. Spaulding, Portland, Oregon, to Patrick D. Goldsworthy, Seattle, August 11, 1966: all in Box 8, Folder: Kennecott Mine General Correspondence 1966, N3CR.
11. A. W. Greeley, Washington, D.C., to Patrick D. Goldsworthy, Seattle, September 2, 1966, Box 8, Folder: Kennecott Mine General Correspondence 1966, N3CR.
12. A. W. Greeley, Washington, D.C., to Lloyd Meeds, Washington, D.C., September 12, 1966, Box 8, Folder: Kennecott Mine General Correspondence 1966, N3CR.
13. Forest Supervisor, Mt. Baker, Bellingham, Washington, to Regional Forester, September 26, 1966, Box: Region 6, Division of Recreation and Lands Recreation Studies (095-76B2307 Box 3-1990), Folder: Wildernesses and Primitive Areas Glacier Peak Wilderness, Folder #3B Mineral Prospecting and Mining, RUSFS; Carlson to Heaton, October 6, 1966.
14. Milvey M. Suchy to the Record, November 7, 1966, Box: Region 6, Division of Recreation and Lands Recreation Studies (095-76B2307 Box 3-1990), Folder: Wildernesses and Primitive Areas Glacier Peak Wilderness, Folder #3B Mineral Prospecting and Mining, RUSFS.
15. Suchy to the Record, November 7, 1966.
16. Quoted in Wilkinson, *Crossing the Next Meridian*, 128. For context, see Miller, *Gifford Pinchot*. One of Grant McConnell's early insights was that the conservation movement more or less *was* Progressivism: "To an important degree, then, the conservation movement of the first part of the century was Progressivism itself. Within a few years it was made to comprehend very nearly all of the specific aims of the Progressive cause." McConnell, "Conservation Movement," 467.
17. Steen, *U.S. Forest Service*, 107, notes that sales, along with grazing permits and recreation leases, always totaled less than annual expenditures in the early years. Neil, *To the White Clouds*, 13 (and throughout) discusses this quick evaporation of opposition once local Idahoans recognized the benefits they received from public lands. Generally, see Hurst, *Law and the Conditions of Freedom*; Klyza, *Who Controls Public Lands;* and Wilkinson, *Crossing the Next Meridian.*
18. McConnell, *Private Power and American Democracy*, 196–245, quotation from 245. A decade earlier McConnell explored the problems of interest groups in the conservation movement along material and nonmaterial lines (or conservative and liberal lines), which had weakened the movement's political effectiveness. See McConnell, "Conservation Movement." "Ideology of decentralization" is from Kaufman, *Forest Ranger*, 83–86. For the battle over the Agnes Creek drainage

amid the larger North Cascades wilderness efforts, including McConnell's role in it, see Marsh, *Drawing Lines*, 41–44.

19. Klyza, *Who Controls Public Lands*, 75.
20. This is not to suggest that large timber companies had received no favors before 1945, but the scale of private cutting from public lands accelerated greatly. See Hirt, *Conspiracy of Optimism*, for the most thorough and searching account of postwar national forest management.
21. Wilkinson, *Crossing the Next Meridian*, 137.
22. Grazing fees were an especially controversial measure the Forest Service introduced in 1906, one that ranchers legally challenged (unsuccessfully) and still bristle against. See Merrill, *Public Lands and Political Meaning*, 63–102; Miller, *Public Lands, Public Debates*, 85–89; Steen, *U.S. Forest Service*, 87–89; Wilkinson, *Crossing the Next Meridian*, 91–92.
23. General Mining Law: "An Act to Provide the Development of the Mining Resources of the United States," *US Statutes at Large* 17 (1872): 91–96, quotation from 91 (emphasis added); Brosnan, "Law and the Environment," 528–30; Gates, *History of Public Land Law*, 723; Klyza, *Who Controls Public Lands?*, 27–47; Wilkinson, *Crossing the Next Meridian*, 28–74.
24. Steen, *U.S. Forest Service*, 296. For the acreages: Hammond, "The Wilderness Act and Mining," 450. See Loop, "Claiming the Cabinets," 50–51, for more on patenting claims in wilderness. The analysis and language of releasing creative energy is the central thread in Hurst, *Law and the Conditions of Freedom*.
25. McConnell, *Private Power and American Democracy*, 3–8, quotation from 3.
26. "Big Mine Near Everett?," *Seattle Post-Intelligencer*, December 23, 1966; "Kennecott Copper Eyes N. Cascades," *Bellingham Herald*, December 22, 1966.
27. William R. Asplund, Wenatchee, Washington, to Patrick D. Goldsworthy, Seattle, October 14, 1966, Box 8, Folder: Kennecott Mine General Correspondence 1966, N3CR.
28. Chriswell, *Memoirs*, 103; Marsh, *Drawing Lines*, 67, 80; Williams, *U.S. Forest Service in Pacific Northwest*, 167, 190–91. Charles Connaughton succeeded Stone in the summer of 1967.
29. "Wilderness Act Bars Pit Mine," *Bellingham Herald*, January 17, 1967.
30. M. Brock Evans, to Herbert J. Stone, Portland, Oregon, January 20, 1967 (Evans mistakenly reversed the J. and Herbert in his name); J. Herbert Stone, Portland, Oregon, to M. Brock Evans, Seattle, January 26, 1967: both in Box 8, Folder: Kennecott Mine General Correspondence 1967, N3CR. Details of Evans's appointment can be found in Brock Evans, Seattle, to Mike [McCloskey], March 1, 1967, Carton 6, Folder 12, GMP.
31. Patrick D. Goldsworthy, Seattle, to Orville L. Freeman, Washington, D.C., March 9, 1967, Box 8, Folder: Kennecott Mine General Correspondence 1967, N3CR.
32. Lloyd Meeds, Washington, D.C., to Arthur W. Greeley, Washington, D.C., January 23, 1967; A. W. Greeley, Washington, D.C., to Lloyd Meeds, February 10, 1967; J. Herbert Stone, Portland, Oregon, to M. Brock Evans, Seattle, January 26, 1967: all in Box 8, Folder: Kennecott Mine General Correspondence 1967, N3CR.

33. According to the best study on forest rangers, such independence was rare. The agency, despite its decentralizing structure, maintained a strong hierarchy. Kaufman, *Forest Ranger*, 4–5, states this clearly, but the entire book explores these tendencies.
34. Forest Supervisor, Mt. Baker, to Regional Forester, March 13, 1967, Division of Recreation and Lands Recreation Studies (095-76B2307 Box 3-1990), Folder: Wildernesses and Primitive Areas Glacier Peak Wilderness, Folder #3B Mineral Prospecting and Mining, RUSFS, emphasis added.
35. Marsh, *Drawing Lines*, 21–23, 27; an example of it working can be found in Sowards, "Administrative Trials," 200–201.
36. His son composed a useful biography: Braman, "Braman, James d'Orma (Dorm)."
37. "Stage Setting for an Open Pit Copper Mine," *Seattle Sunday Post-Intelligencer*, August 27, 1967; "Restrictions Announced on Open Copper Mine," *Seattle Post-Intelligencer*, August 27, 1967.
38. "Restrictions Announced on Open Copper Mine"; Wilderness Act. See also the summary provided by Brock Evans in M. Brock Evans to Gene Marine, San Francisco, October 10, 1967, Box 22, Folder 40, BEP.
39. "Ride-In to Glacier Peak," *Seattle: The Pacific Northwest Magazine*, October 1967, 23–27, quotations from 26, 27. Hays, *American People and National Forests*, 92–105, describes some of the ways the Forest Service struggled to respond to its expanding mission.
40. "Ride-In to Glacier Peak."
41. Allan May, "Forest Service Expects Court Test on Mine," *Everett Herald*, August 18, 1967.
42. Paul B. Jessup, New York, to Edward P. Cliff, Washington, D.C., August 28, 1967; Edward P. Cliff, Washington, D.C., to Paul B. Jessup, New York, September 12, 1967: both in Box: Region 6, Division of Recreation and Lands Recreation Studies (095-76B2307 Box 3-1990), Folder: Wildernesses and Primitive Areas Glacier Peak Wilderness, Folder #3B Mineral Prospecting and Mining, RUSFS.
43. Jessup to Cliff, August 28, 1967; Cliff to Jessup, September 12, 1967. In a public address at the American Mining Congress two years later, Cliff advised the industry to pay more attention to its image and urged miners to recognize shifting historical currents that required it to address public opinion more systematically. Cliff, "Mining and the National Forests."
44. "Ride-In to Glacier Peak," 27.
45. Leshy describes this ambiguous situation better than anyone. Leshy, *Mining Law*, 229–42. A useful example is Loop, "Claiming the Cabinets."

CHAPTER 4. THE SUMMIT

1. Patrick D. Goldsworthy, Seattle, to George Marshall, Los Angeles, November 12, 1966; George Marshall, Los Angeles, to C. D. Michaelson, New York, December 2, 1966: both in Carton 46, Folder 3, GMP. Marshall apparently wrote to Frank R. Milliken and C. D. Michaelson on December 2, 1966, and the board of directors on December 5, 1966. All the letters can be found in the same folder.

For the reference to *Wild Cascades*, see George Marshall, Los Angeles, to Patrick D. Goldsworthy, Seattle, December 13, 1966, Box 8, Folder: Kennecott Mine General Correspondence 1966, N3CR. Conservationists coordinated their activities and shared their correspondence. Consequently, copies of the same letters sometimes can be found in several collections. For example, the George Marshall letters cited above are also included in the North Cascades Conservation Council Records. Because citing duplicate collections could become unwieldy, I typically cite what seems closest to the original source. Hence, the citation from the Marshall Papers here.

2. Marshall to Michaelson, December 2, 1966.
3. A. F. Mayne, Montreal, to George Marshall, San Francisco, January 9, 1967 ("mutually constructive" quotation); Ellison L. Hazard, New York, to George Marshall, San Francisco, December 23, 1966; Robert G. Stine, Boston, to George Marshall, San Francisco, December 20, 1966 ("maintain the general area" quotation); Peter Lawson-Johnston, New York, to George Marshall, San Francisco, December 19, 1966; E. L. Steiniger, New York, to George Marshall, San Francisco, December 19, 1966: all in Box 8, Folder: Kennecott Mine General Correspondence 1966, N3CR.
4. Frank R. Milliken, New York, to George Marshall, San Francisco, December 15, 1966; and Paul B. Jessup, New York, to Sierra Club Books, San Francisco, December 15, 1966: both in Carton 46, Folder 3, GMP.
5. LeCain, *Mass Destruction*, Kindle edition, location 415.
6. Hawley, *Kennecott Story*, 2.
7. Initially Kennecott was a holding company for all the Guggenheim family's copper properties, and the Utah Copper Company, which started operations at Bingham Canyon, continued to operate under its own name. Utah Copper Company became a wholly owned Kennecott subsidiary in 1936. Details of Kennecott's formation and development are from Arrington and Hansen, *Richest Hole on Earth*, 67–69, 84–86; Hawley, *Kennecott Story*.
8. One set of figures shows the mine yielding just shy of 2 percent in 1905. Ten years later the percentage had dropped to 1.43 percent. It dropped below 1 percent in 1927. By the mid-1950s, it was below 0.9 percent. Arrington and Hansen, *Richest Hole on Earth*, 90–91.
9. Arrington and Hansen, *Richest Hole on Earth*, 52–54, for what the authors call "opencut" mining; LeCain, *Mass Destruction*, Kindle edition (ore concentration from location 1545, power use from location 434, quotations from location 217 and 114, start date from location 1506, more context in introduction and chapter 4). LeCain's account of the "envirotechnical" system open-pit mining produced emphasized the ways that system helped put distance between the ecological costs of production and consumption. Mass destruction thus joined mass production (think Henry Ford's assembly line) and mass consumption (think chain stores and standardized products) in a knot of economic and ecological consequences of modern industrialism. The distance between sites of extraction, production, and consumption allowed Americans to ignore more easily the costs of twentieth-century consumerism in what it exacted from the earth, not

to mention from laborers. LeCain's book is influenced by Cronon, *Nature's Metropolis*.

10. Increasing economies of scale are seen to be positive economic developments when production is raised to a point that allows per unit profit to also increase. This approach generates some economic efficiencies, but the costs rise significantly. Small businesses generally cannot compete at large economies of scale, tending to leave industries controlled by a few companies.
11. Arrington and Hansen, *Richest Hole on Earth*, 70; Hyde, *Copper for America*, xvii, 160–61, 181–85.
12. Hawley, *Kennecott Story*, 265, 267, 274; Hyde, *Copper for America*, 178, 181.
13. Hyde, *Copper for America*, 181–85.
14. Kennecott statistics come from *Walter R. Skinner's "Mining Year Book," 1966*, 336–39; *Walter R. Skinner's Mining Year Book 1969*, 342–45. Northwest conservationists had access to this data and shared it among themselves. For instance, photocopies of the 1966 yearbook appear in Box 8, Folder: Kennecott Mine—Facts and Statistics 1967–1971, N3CR. The cash Kennecott had on hand was risky, so the company acquired Peabody Coal Company, the largest in the country, at an inflated price. This acquisition turned out to be a problem because of Peabody's aging equipment and the Federal Trade Commission's finding that the acquisition violated competition rules. Kennecott sold Peabody for a profit, but the episode was a costly distraction. See Hawley, *Kennecott Story*, 293–94.
15. General Services Administration, "Brief History of GSA."
16. GSA News Release, May 9, 1966 (quotations); and GSA News Release, August 2, 1966: both in Box: Region 6, Division of Recreation and Lands Recreation Studies (095-74E0241 Box 8/147765), Folder: Wildernesses and Primitive Areas Glacier Peak Wilderness, Folder 3a), RUSFS.
17. The Department of Interior worked with mineral claims but did not have jurisdiction over national forests.
18. John G. Harlan Jr., Washington, D.C., to Paul R. Ignatius, Washington, D.C., March 31, 1967, Box: Region 6, Division of Recreation and Lands Recreation Studies (095-74E0241 Box 8/147765), Folder: Wildernesses and Primitive Areas Glacier Peak Wilderness, Folder 3a), RUSFS.
19. Harlan to Ignatius; Paul R. Ignatius, Washington, D.C., to Henry M. Jackson, Washington, D.C., April 7, 1967 (first quotation); Milvoy M. Suchy, Portland, Oregon, to the Record, April 3, 1967 (second quotation): both in Box: Region 6, Division of Recreation and Lands Recreation Studies (095-74E0241 Box 8/147765), Folder: Wildernesses and Primitive Areas Glacier Peak Wilderness, Folder 3a), RUSFS.
20. "Big Mine Near Everett?," *Seattle Post-Intelligencer*, December 23, 1966; "Huge Open Copper Mine Considered for Cascades," *Bremerton (WA) Sun*, December 23, 1966.
21. Brock Evans, Seattle, to Governor [Dan] Evans, January 7, 1966, Box 22, Folder 40, BEP.
22. Goldsworthy's notes from the meeting are in Box 8, Folder: Kennecott Mine—Minutes of Meetings with Kennecott Copper Corporation 1969, n.d., N3CR,

quotations from 2 and 3. These notes were closely formulated into an article that appeared as Goldsworthy, "Kennecott Meets with Conservation Leaders," and in *Mountaineer*, April 1967, 11–12. The public relations officer appears in Goldsworthy's notes only as "Mr. Hayes." No first name was given, and I have located no other instance where he appears in the historical record.

23. The copper production estimate comes from Paul B. Jessup, New York, to George Alderson, Washington, D.C., February 28, 1967, Box 8, Folder: Kennecott Mine General Correspondence 1967, N3CR. All other figures are from Goldsworthy's notes.
24. Goldsworthy's notes.
25. Goldsworthy's notes.
26. Paul B. Jessup, New York, to George Alderson, Washington, D.C., February 22, 1967, Box 8, Folder: Kennecott Mine General Correspondence 1967, N3CR.
27. Goldsworthy notes.
28. Goldsworthy notes, emphasis in original. A letter from a Kennecott official soon after the summit revealed the company's strategy to minimize effects through "adoption of reasonable restoration procedures when operations terminate." Jessup to Alderson, February 22, 1967.
29. For those two other campaigns specifically, see Pearson, *Still the Wild River Runs;* Schrepfer, *Fight to Save the Redwoods*. For the place of those campaigns in the Sierra Club's history, see Cohen, *History of the Sierra Club*, 299–323.
30. Goldsworthy notes.
31. George Marshall, Los Angeles, to Frank R. Milliken, New York, February 14, 1967, Carton 6, Folder 20, GMP.
32. Patrick D. Goldsworthy, Seattle, to F. R. Milliken, New York, February 14, 1967, Carton 46, Folder 4, GMP.

CHAPTER 5. THE SECRETARY

1. Marsh, *Drawing Lines*, 82–83, 106, demonstrates how Secretary Freeman proved unmoved by citizen appeals to protect wilderness in Washington's Alpine Lakes and Oregon's French Pete areas while encouraging timber sales to proceed without further delays.
2. "Livermore Calls Proposed Sierra Road 'Tragedy,'" *San Francisco Chronicle*, April 9, 1967.
3. Louter, *Windshield Wilderness*.
4. Cohen, *History of the Sierra Club*, 121–34, covers these conferences.
5. Cohen, *History of Sierra Club*, 120.
6. Cohen, *History of Sierra Club*, 121, 123; Wyss, *Man Who Built the Sierra Club*, 78. One of Brower's biographers indicates that Brower credited the conference idea to Benton MacKaye, a Wilderness Society founder, but nowhere else does that claim appear, and the claim is not cited. Turner, *David Brower*, 61.
7. Cohen, *History of Sierra Club*, 124–26, 131–34.
8. Brower, *For Earth's Sake*, 270.

9. At the same time the conferences began in 1949, Brower changed the tone of the *Sierra Club Bulletin* that he took over editing in 1946. Rather than just report on outings, the *Bulletin* started publishing critical pieces about policy and reprinting articles about wilderness from the likes of Bob Marshall and Bernard DeVoto. Cohen, *History of Sierra Club*, 123.
10. Sutter's *Driven Wild* best explores these founders' ideas.
11. Wyss, *Man Who Built the Sierra Club*, 79. See also Turner, *David Brower*, 84.
12. Turner, *David Brower*, 62, 84.
13. Harvey, *Wilderness Forever*, 136–37, 170.
14. Turner, *David Brower*, 62. Details of Douglas's 1961 talk are found in Sowards, *Environmental Justice*, 73–76.
15. Harvey, *Forever Wild*, 190.
16. David Stout, "Orville Freeman, 60's Agriculture Secretary, Dies at 84," *New York Times*, February 22, 2003; "Orville L. Freeman, 84: U.S. Secretary of Agriculture under Kennedy, Johnson," *Los Angeles Times*, February 22, 2003.
17. *Hearing before the Committee on Agriculture and Forestry*, 31–32.
18. Harvey, *Wilderness Forever*, 219; Scott, *Enduring Wilderness*, location 700–705.
19. Michael McCloskey, *In the Thick of It*, 29. See also Hirt, *Conspiracy of Optimism*, 226–27.
20. Marsh, *Drawing Lines*, 72, 82–83, 106.
21. "Orville L. Freeman, 84."
22. Freeman, "Address," 107–15.
23. Freeman, "Address," 109–10.
24. The apparent uniqueness of the nation's wild spaces helped generate political momentum to preserve places like Yellowstone in the nineteenth century as Americans sought ways to develop and express nationalist ideologies and identities. This is the argument of some of the earliest environmental history of preservation. See Nash, *Wilderness and the American Mind*, esp. 67–83; Runte, *National Parks*, esp. 11–27.
25. This is a major point, often criticized and misunderstood, in Cronon, "Trouble with Wilderness."
26. See, for example, Brinkley, *Wilderness Warrior*, 239–47; Cronon, "Trouble with Wilderness."
27. The style and effectiveness of Freeman's address might be compared with chief forester Ed Cliff's address in the morning. Cliff spoke almost solely in policy terms, rarely mentioning specific places and never suggesting the value of beauty. See Cliff, "Wilderness Act and the National Forests," 6–12,
28. Freeman, "Address," 110; Stout, "Orville Freeman" (on Freeman being a deacon). Freeman used his role as a Lutheran church deacon to help blunt anti-Catholicism against John F. Kennedy in the presidential campaign, which may have helped gain Freeman's appointment.
29. Muir, *Yosemite*, 261. For an interpretation of Muir's deep spiritual values, see Sowards, "Spiritual Egalitarianism."
30. Freeman, "Address," 110–11, quotation from 111.
31. Freeman, "Address," 110, 111.

32. Freeman, "Address," 110, 115.
33. Brock Evans thanked Freeman in M. Brock Evans, Seattle, to Orville H. Freeman, Washington, D.C., April 21, 1967, Box 22, Folder 40, BEP. The president of the Sierra Club wrote to the president of Kennecott to inform him of the secretary's position in George Marshall, Los Angeles, to Frank R. Milliken, New York, April 28, 1967, Box 22, Folder 39, BEP. And the United Steelworkers of America issued a press release critical of Kennecott's position in a "News from the USWA" press release, April 7, 1967, Box 22, Folder 45, BEP.
34. Dayton, "Behind the By-Lines."
35. The month before the speech, Patrick D. Goldsworthy wrote to Freeman and emphasized that this was the first test of the Wilderness Act. See Goldsworthy, Seattle, to Orville L. Freeman, Washington, D.C., March 9, 1967, Box 8, Folder Kennecott Mine General Correspondence 1967, N3CR.

CHAPTER 6. THE DOCTOR

1. A photo of Darvill with the painting appeared in "Copper Firm Stockholder Hits Open-Pit Mine Plan," an unsourced newspaper clipping in Box 22, Folder 49, BEP.
2. The details of Darvill's life come from his obituary, "Fred T. Darvill M.D.," http://www.legacy.com/obituaries/skagitvalleyherald/obituary.aspx?n=fred-t-darvill&pid=100580180&. On Darvill's trip, see the entry for Fred T. Darvill Papers, University of Utah Libraries, Special Collections, http://archiveswest.orbiscascade.org/ark:/80444/xv05230/pdf. On Glen Canyon, see Farmer, *Glen Canyon Dammed*, for specifics of Glen Canyon; for larger context, see Harvey, *Symbol of Wilderness*.
3. "Open-Pit Mine in Cascades Draws Fire," *Skagit Valley Herald*, December 23, 1966; F. T. Darvill, "Deplores Open Pit Mine Proposal in Wilderness," *Skagit Valley Herald*, December 28, 1966; Henry M. Jackson, Washington, D.C., to Lloyd Seabury, Anacortes, Washington, January 6, 1967; Box 8, Folder: Kennecott Mine General Correspondence 1967, N3CR.
4. Needham, *Power Lines*, 185–212. Pearson argues that pragmatic politics might have tipped the balance in the decisions not to build the two downriver dams but does not dismiss the power of grassroots organizing. See Pearson, *Still the Wild River Runs*. Needham's account emphasizes the energy trade-off that arose because of the elimination of the Grand Canyon dams, namely the coal-mining and building of power plants on the Navajo Reservation. These developments fueled growth in Phoenix, provided jobs on the reservation, dirtied the air, and harmed the health of the region's residents. The other priority for the Sierra Club was working to establish the Redwood National Park in California. Cohen, *History of the Sierra Club*, 388.
5. As discussed later in the book, the high price of copper abroad partly explains the putative copper shortage in the United States. Calls for copper production for the war effort, then, masked some of the market mechanisms at work. Lloyd Meeds, Washington, D.C., to F. T. Darvill, Mount Vernon, Washington, December 29, 1966, Box 8, Folder: Kennecott Mine General Correspondence 1966, N3CR.

6. "Copper Firm Stockholder Hits Open-Pit Mine Plan."
7. F. T. Darvill, Mount Vernon, Washington, to Kennicott [*sic*] Copper Company, New York, January 19, 1967, Box 8, Folder: Kennecott Mine General Correspondence 1967, N3CR.
8. Darvill to Kennicott [*sic*], January 19, 1967.
9. Frank R. Milliken, New York, to F. T. Darvill, Mount Vernon, Washington, February 6, 1967; F. T. Darvill, Mount Vernon, Washington, to George F. Joffe, New York, February 8, 1967: both in Box 22, Folder 38, BEP.
10. Adam Rome summarizes the ideas and activities of postwar liberals in *Genius of Earth Day*, 10–20.
11. Michael McCloskey, "Can Recreational Conservationists Provide for a Mining Industry?," 66–67 (quotations), 69.
12. Cohen, *Consumers' Republic*, esp. 7–15, 345–97.
13. One oft-cited example in western history is the Mexican-American Community Services Organization (CSO), which started in Los Angeles in the 1940s and did much to support the community's political, educational, and legal needs. The lead organizer, Fred Ross, knew Alinsky and received support from him. The CSO thus had an imprint of Alinsky even though Ross started organizing in Los Angeles before meeting him. Ross hired and mentored Cesar Chavez in the CSO, so it is often reported that that there is a line of Alinsky-inspired organizing among Mexican-American activists. See Gutierrez, *Walls and Mirrors*, 168–69; Shorris, *Latinos*, 305.
14. Alinsky, *Reveille for Radicals*, by-laws on 221–28.
15. The need for a new strategy arose because of the breakdown of negotiations. FIGHT pushed Kodak to hire more African Americans. Senior executives reached an agreement with activists before other senior executives the following day publicly reneged on the deal. FIGHT, Alinsky, and others sought a way to engage a broader constituency to apply pressure on the corporation. See Horwitt, *Let Them Call Me Rebel*, 489–95; Finks, *Radical Vision of Saul Alinsky*, 201–11.
16. Alinsky, *Rules for Radicals*, 165–83; Finks, *Radical Vision*, 217–22; Horwitt, *Let Them Call Me Radical*, 495–502. Darvill's vice presidency is mentioned in "Copper Firm Warned."
17. "Gilbert Goes to Meetings and Raises Roof," *Newsweek*, March 13, 1937, 27.
18. M. J. Rossant, "All You Need Is One Share," *New York Times*, May 7, 1967; Leonard Sloane, "Lewis Gilbert, 86, Advocate of Shareholder Rights," *New York Times*, December 8, 1993; Gilbert, *Dividends and Democracy*, ix ("fists revolving"), 18–32 (on angering board members), x ("deeply conservative").
19. Rossant, "All You Need Is One Share."
20. Rossant, "All You Need Is One Share."
21. The classic account of the rise of environmentalism as a movement with a focus on issues related to consumption is Hays, *Beauty, Health, and Permanence*, esp. 13–39.
22. F. T. Darvill, Mount Vernon, Washington, to J. Michael McCloskey, San Francisco, March 24, 1967, Box 22, folder 38, BEP.

23. Brock Evans, Seattle, to editor [*Medical World News*], New York, April 24, 1967, Box 22, Folder 40, BEP.
24. Fred T. Darvill, Mount Vernon, Washington, to Pat[rick Goldsworthy], February 8, 1967, Box 22, folder 38, BEP.
25. George Marshall, Los Angeles, to Frank R. Milliken and Board of Directors [Kennecott], New York, April 28, 1967, Carton 6, Folder 20, GMP.
26. "Copper Firm Stockholder Hits Open-Pit Mine Plan" (impassioned talk); Malcolm R. Wilkey, New York, to Fred T. Darvill, Jr., M.D., Mount Vernon, Washington, January 23, 1967 (details of time and place of meeting); and "Statement of Fred Darvill, MD, Kennecott Copper Corporation, Annual Stockholders' Meeting, Biltmore Hotel, May 2, 1967" (all other quotations): the latter two both in Box 8, Folder: Kennecott Mine—Speeches and Writings, misc. 1960–1970, N3CR.
27. Statement of Fred T. Darvill.
28. Kennecott Copper Corporation, "Summary Report, Annual Meeting of Stockholders, Held May 2, 1967," Box 8, folder: Kennecott Mine—Miscellany 1964–1973, N3CR; "Kennecott Profit Showed Big Gain in First Quarter," *Wall Street Journal*, May 3, 1967.
29. Walt Woodward, "Copper Firm Warned against Cascades Mine," *Seattle Times*, undated clipping in North Cascades Conservation Council Records, Special Collections, University of Washington, Seattle.
30. "Kennecott Profit Showed Big Gain"; Kennecott, "Summary Report." Mrs. Weinstein's first name is not included in the sources. See also Robert A. Wright, "Copper Concerns Report Earnings," *New York Times*, May 3, 1967; "Kennecott Hears Protest to N. Cascades Mining," *Oregonian*, May 3, 1967; "Copper Firm Directors Hear Conservation Plea," *Columbian* (Vancouver, WA), May 3, 1967.
31. Woodward, "Copper Firm Warned against Cascades Mine."
32. Kennecott, "Summary Report, Annual Meeting of Stockholders."
33. Woodward, "Copper Firm Warned against Cascades Mine."
34. "Crisis at Miners Ridge," *Washington Post*, May 3, 1967 (all quotations); Eric Wentworth, "U.S. Forest Service Strives to Prevent Big Open-Pit Mine," *Washington Post*, May 2, 1967.

CHAPTER 7. THE JUSTICE

1. "Stop Kennecott Group Organized," *Wenatchee (WA) Daily World*, April 30, 1967; Rob and Cindy Cole, "An Open Pit on Miner's Ridge Large Enough to Be Seen on the Moon!," unsourced news clipping in Box 22, Folder 49, BEP.
2. Harvey, *Wilderness Forever*, 175.
3. This is the argument about Douglas generally in Sowards, *Environmental Justice*.
4. Fox, *American Conservation Movement*, 239, calls Douglas the "most prominent conservationist in public life" in the three decades after 1945.
5. Two thorough biographies are Simon, *Independent Journey*; Murphy, *Wild Bill*.
6. A good example of the Douglas-as-outdoorsman trope is in Neuberger, "Mr. Justice Douglas," in *They Never Go Back to Pocatello*, 108–25 (first published in

Harper's, August 1942). His own memoir solidified this reputation. Douglas, *Of Men and Mountains*.

7. Sowards, *Environmental Justice*, 31–57.
8. See Douglas, *My Wilderness: The Pacific West;* Douglas, *My Wilderness: East to Katahdin*.
9. Sowards, *Environmental Justice*, 58–80.
10. Douglas, *My Wilderness: The Pacific West*, 156, 155, 167.
11. Douglas, *My Wilderness: The Pacific West*, 159, 158, 167.
12. Douglas, *My Wilderness: The Pacific West*, 165.
13. Sowards, *Environmental Justice*, 48–56, contains a full account of the Olympic Beach hike, complete with citations. The account here follows that one. For more on Dyer, see Arntz, *Extraordinary Women Conservationists*, 29–31, 37–44.
14. Louter, *Windshield Wilderness*.
15. Conservationists from their beginning faced challenges over gender, such as a complaint referring to Hetch Hetchy defenders as "short-haired women and long-haired men." Quoted in Unger, *Beyond Nature's Housekeepers*, 100. See Harvey, *Wilderness Forever*, 75, for "flower picker."
16. Warren, *Hunter's Game*.
17. Merrill makes this case concerning grazing lands in *Public Lands and Political Meaning*, 7.
18. Perhaps best articulated in Douglas, *Farewell to Texas*.
19. Marjorie Jones, "Douglas Leads Glacier Peak Protest," *Seattle Times*, August 6, 1967.
20. Poehlman, *Darrington*, 165–70, for information on mills.
21. Jones, "Douglas Leads Glacier Peak Protest."
22. Maribeth Morris, "Protestors Crash Douglas Camp-In," *Seattle Post-Intelligencer*, August 6, 1967, 3.
23. Jones, "Douglas Leads Glacier Peak Protest"
24. White, "'Are You an Environmentalist?,'" 171–85. The essay title came from a bumper sticker found in Northwest timber towns in the 1990s.
25. Historian Thomas Andrews wrote about a laborer in the Colorado coalfields who offered a similar statement about the value of work in transforming landscapes. See Andrews, "Made by Toile?"
26. Morris, "Protestors Crash Douglas Camp-In." Neidigh used the exact same "shot in the arm" phrase, suggesting literally a common script. See Jones, "Douglas Leads Glacier Peak Protest."
27. Nye, *America as Second Creation*. Klingle explores this in an urban setting in *Emerald City*, 86–118.
28. "Justice Douglas Leads Hikers to Glacier Peak," *Wenatchee (WA) Daily World*, August 7, 1967.
29. M. Brock Evans, to Gene Marine, San Francisco, October 10, 1967, in Brock Evans Papers, Special Collections, University of Washington, Seattle.
30. Robert Michael Pyle, personal communication to the author, August 7, 2015.
31. Pyle recalled Douglas listening to protesters.
32. Pyle, personal communication.

33. Jones, "Douglas Leads Glacier Peak Protest."
34. Pyle, personal communication. See also Pyle, "Reflections on 50 Years of Engagement."
35. For more on the scene, see "Miner's Ridge Film: Notes to Editor-Producer," May 13, 1968," Box 8, Folder: Kennecott (Mine) Film 1968, N3CR; "Justice Douglas Leads Hikers"; M. Brock Evans to Dick Miller, Pittsburgh, October 10, 1967, Box 22, Folder 40, BEP. Although Douglas's chapter in *My Wilderness: The Pacific West* demonstrated that he knew of the potential for a copper mine, he was first alerted to the *imminent* threat when the president of the North Cascades Conservation Council wrote to the secretary of agriculture and copied Douglas. See Patrick D. Goldsworthy, Seattle, to Orville L. Freeman, Washington, DC, March 9, 1967, Box 8, Folder: Kennecott Mine General Correspondence 1967, N3CR.
36. Pyle, personal communication. No one recorded Douglas's full remarks the day of the protest. The Sierra Club planned a film about Kennecott and so wanted to capture Douglas's sentiments. The following summer, North Cascades Conservation Council president Patrick Goldsworthy and student-activist Benjamin Shaine recorded Douglas, who spoke as though he were standing on the banks of the Suiattle, but he was at his home in Goose Prairie, Washington, near Mount Rainier National Park. The North Cascades Conservation Council records contain two folders of materials related to the film; rather than cite the voluminous correspondence, I refer readers to Box 8, Folder: Kennecott (Mine) Film 1967–1968, N3CR. See also Benjamin A. Shaine, "Kennecott on Miners Ridge: A Description and Analysis of the Controversy over the Proposal of the Kennecott Copper Corporation to Construct an Open-Pit Copper Mine in the Glacier Peak Wilderness Area of Washington," honors thesis, Oberlin College, 1969, Box 8, Folder: Kennecott Mine—Speeches and Writings—"Kennecott on Miner's Ridge" by Benjamin Shaine, unpaginated, N3CR. 37. Quoted in Morris, "Protestors Crash Douglas Camp-In."
38. For more on the way Douglas used his conservation thinking in writing legal opinions, see Sowards, *Environmental Justice*, 112–37.
39. "Remarks of Justice William O. Douglas regarding Kennacott [*sic*] Copper Corp. Proposed Open-Pit Copper Mine on Miners Ridge, July 21, 1968," Box 8, Folder: Kennecott (Mine) Film 1967–1968, N3CR.
40. "Remarks of Justice William O. Douglas."
41. Sowards, *Environmental Justice*, 71–80.

CHAPTER 8. THE PARK

1. Walt Woodward, Seattle, to Brian Corcoran, Washington, D.C., September 19, 1968, Box 210, Folder 7, HMJP.
2. Quoted in Manning and North Cascades Conservation Council, *Wilderness Alps*, 143, emphasis in original.
3. Henry M. Jackson, Washington, D.C., to Paul Brooks, Boston, April 24, 1967; Box 209, Folder 27, HMJP.
4. By some accounts, Brower gave up on the Forest Service after frustrating experiences in California's eastern Sierra Nevada. Wyss, *Man Who Built the Sierra Club*,

90–106. Simons died in 1960, aged twenty-five, after contracting a rare form of hepatitis. Even in his short lifetime he contributed importantly to Northwest conservation. See Wyss, *Man Who Built the Sierra Club*, 112–14, on Simons's work.

5. David R. Brower, "Conservation: Mountain Region," *New York Times*, February 1, 1959.
6. John B. Oakes, "Conservation: Parks or Forests?," *New York Times*, February 7, 1960.
7. A copy of Pelly's resolution is available in Carton 78, Folder 16, Sierra Club Records. Bancroft Library, University of California, Berkeley.
8. Michael McCloskey, *In the Thick of It*, 43; Danner, *Crown Jewel Wilderness*, 132–34.
9. North Cascades Study Team, *North Cascades Study Report: A Report to the Secretary of the Interior and the Secretary of Agriculture* (correspondence between the cabinet secretaries and the president found on 153–55, public interest quotation from ii). Danner, *Crown Jewel Wilderness*, 123–47, and Marsh, *Drawing Lines*, 52–60, provide excellent context and details on this era.
10. George Marshall, Los Angeles, to Harvey Manning, Issaquah, Washington, October 30, 1961, Carton 5, Folder 27, GMP.
11. The "openness" of public lands often has been an ideal, not a reality. Racial minorities have been excluded by law or by practice. See Finney, *Black Faces, White Spaces*. Also, economic activity could restrict access, at least temporarily, such as during a timber harvest.
12. North Cascades Study Team, *North Cascades Study Report*, 79, 82. The study team also reported on several petitions it had received, dating back to 1960, with thousands of signatures. Approximately twenty-three thousand signatures were attached to petitions asking for a study of national park potential. Not quite two thousand favored continued USFS management. About 10 percent of those who testified submitted statements on behalf of organizations, such as chambers of commerce, local governments and schools, and commodity groups, all of whom generally favored multiple use and opposed a park because of fears of lost revenue. About 20 percent of those who testified represented conservation groups that favored the national park. North Cascades Study Team, *North Cascades Study Report*, 83–84.
13. North Cascades Study Team, *North Cascades Study Report* (summarized recommendations are on pp. 14–17, details on the wilderness area are on pp. 88–89, and details on the proposed national park are on pp. 90–109). Mining appears negligibly in the report, which ended with a series of dissenting opinions. Stratton made a number of recommendations slightly at odds with the team's conclusions and concluded with a prescient statement about the Glacier Peak Wilderness Area. He believed it too fragile to be included in the national park and to endure the mass recreational use the team anticipated in a park. Looking to the future, he imagined that wilderness areas would require management to a greater degree than they had before. Stratton could easily imagine wilderness users destroying the very thing they valued with overuse. He anticipated the need to ration wilderness experiences much as one controlled hunting mountain goats. Glacier Peak

would likely be an early casualty because of its easy access from Seattle. Although Stratton put it in different terms, these concerns were familiar to many, including Justice Douglas, who initially opposed the national park on the grounds of over-development. North Cascades Study Team, *North Cascades Study Report*, 127–29; William O. Douglas to George B Hartzog Jr., Washington, D.C., March 9, 1967, Box 558, Folder: North Cascade Mountains (1967–1968), WODP.

14. Strangely, Senator Jackson invited a former superintendent at Olympic National Park to the event, Fred J. Overly, who during his park superintendency had overseen clearcutting in the park in clear violation of the law. In 1966, working as regional director of the Bureau of Recreation, Overly was still a controversial and distracting figure. He continued to advocate cuts from Olympic National Park, and his presence at this press conference generated the distinct impression that creating North Cascades National Park required a reduction in size of the Olympic National Park. The hearings Jackson would hold in February, before the Committee on Interior and Insular Affairs, specifically linked the two parks despite the fact that the study team's report was not charged with examining Olympic National Park and despite the fact that its only mention came once in extended remarks by George Selke. To casual observers, Olympic National Park was beside the point, but insiders worried about Overly's history and Jackson's commitment. Manning and North Cascades Conservation Council, *Wilderness Alps*, 179–91, makes the concern of locals about Olympic clear.
15. *Hearings before the Committee on Interior and Insular Affairs, United States Senate, Study Team Report of the Recreational Opportunities in the State of Washington* (hereafter *Hearings, Study Team Report*), 450 ("national tradition"), 451 ("greatest good"), 650 ("more enlightened definition").
16. *Hearings, Study Team Report*, 650 ("least visible" and "most valuable"), 451 ("concealed or gradually healed").
17. *Hearings, Study Team Report*, 533.
18. "Copper Exports Will Be Curbed," *New York Times*, January 12, 1966.
19. See Kaufman, *Henry M. Jackson*. Jackson, for all his liberalism on environmental matters, remained a steadfast hawk on defense matters, often referred to as "the senator from Boeing" for his ability to ensure defense contracts for the Seattle-based corporation.
20. *Hearings, Study Team Report*, 651.
21. See, for example, *Hearings, Study Team Report*, 87–89, 217–19.
22. *Hearings, Study Team Report*, 227.
23. Johnson, "Special Message to Congress."
24. *Hearings before the Subcommittee on Parks and Recreation*, 5.
25. Robert M. Pyle, Seattle, to Senator Henry M. Jackson, Washington, D.C., May 9, 1967, Box 210, Folder 1, HMJP.
26. The figures are derived from my analysis of the testimony in *Hearings before the Subcommittee of Parks and Recreation*.
27. Michael McCloskey, *In the Thick of It*, esp. 40–45.
28. Hammond, "Wilderness Act and Mining," 449.
29. *Hearings before the Subcommittee on Parks and Recreation*, 171, emphasis added.

30. Arntz, *Extraordinary Women Conservationists*, 29–35, 37–44. Arntz's book is not always reliable on specific details, but it is one of the only attempts to include a sustained focus on Dyer. See also Dyer's appearance in Marsh, *Drawing Lines*, 42, 55, 84; Sowards, *Environmental Justice*, 49–50, 54. Perhaps Dyer's greatest claim to fame—although few know this—is her use of the word "untrammeled" to describe wilderness, a descriptor Howard Zahniser adopted in his drafting of the Wilderness Act. See Harvey, *Wilderness Forever*, 119, 202.
31. *Hearings before the Subcommittee on Parks and Recreation*, 457.
32. See, for example, *Hearings before the Subcommittee on Parks and Recreation*, 201, 334, 677, 733, 879, 891, 907.
33. *Hearings before the Subcommittee on Parks and Recreation*, 99 ("Obsolete mining laws"), 179 (Avery). Avery's first name was misspelled as "Abbigail" in the published record. See also *Hearings before the Subcommittee on Parks and Recreation*, 179, 211, 322, 733.
34. Darvill, *Hiking the North Cascades*, 183–85.
35. *Hearings before the Subcommittee on Parks and Recreation*, 74, 383, 795, 899 (quotation), 973. The Environmental Protection Agency continues to monitor the Holden cleanup site (see https://www.epa.gov/enforcement/case-summary-epa-and-forest-service-issue-uao-holden-mine-site), and the company in charge of the Holden cleanup, Rio Tinto, is the same company that would acquire Kennecott Copper. See Rio Tinto's page on Holden's cleanup, http://www.holdenminecleanup.com/.
36. Douglas, "America's Vanishing Wilderness"; Sowards, *Environmental Justice*, 2–3, 81–111.
37. *Hearings before the Subcommittee on Parks and Recreation*, 311–13, 397, 952 ("canker sore"). Some did see tourist possibilities like those developed in Utah. See *Hearings before the Subcommittee on Parks and Recreation*, 569, 793. For more on the Bingham Pit's destructiveness, see LeCain, *Mass Destruction*.
38. *Hearings before the Subcommittee on Parks and Recreation*, 155, 397, 1070.
39. One teenager shared a particularly feisty testimony: "I am not going to stand by idly and let our state's future be guided by the lustful hands of commercial investors and industrialists." *Hearings before the Subcommittee on Parks and Recreation*, 845–46. Several soldiers from Vietnam wrote, including one who said that his "pinups" were not of "lovely young ladies" but landscape shots of the North Cascades. *Hearings before the Subcommittee on Parks and Recreation*, 1067.
40. See, for example, *Hearings before the Subcommittee on Parks and Recreation*, 394, 795, 814 (quotation), 886, 955, 960, 991.
41. Rape references came up often in the testimony. See, for example, *Hearings before the Subcommittee on Parks and Recreation*, 798, 842, 857, 1005.
42. William P. Jeske, Seattle, to Henry M. Jackson, Washington, D.C., October 30, 1967, Box 209, Folder 29, HMJP.
43. Evelyn Gayman, Los Angeles, to Lyndon B. Johnson, Washington, D.C., July 17, 1967; Henry M. Jackson, Los Angeles, to George Murphy, Washington, D.C., August 4, 1967: both in Box 209, Folder 28, HMJP.
44. The North Cascades Conservation Council published the 1937 proposal in their newsletter *Wild Cascades*, February 1963.

45. Sierra Club, "A North Cascades National Park for America's Alps," Box 2, folder: Washington—North Cascades, SCNO.
46. See, for example, Connelly, "North Cascades Conservation Council." An example of the ad can be found in Carton 46, folder 9, GMP. The historical record is a bit murky. I have not seen the ad in the *New York Times* or the *Los Angeles Times* itself, nor did I locate any specifics about it in the Sierra Club records. However, the ad exists in several archival collections—just not clearly from either the *New York Times* or *Los Angeles Times.*
47. *Wild Cascades*, December 1966–January 1967.
48. Ads like this one, but for the Grand Canyon, caused the Sierra Club problems, eventually pushing the Internal Revenue Service to revoke the club's tax-exempt status. See Turner, *David Brower*, 122–26; Wyss, *Man Who Built the Sierra Club*, 208–14. Michael McCloskey, Brower's successor, criticized this strategy in *In the Thick of It*, 94. When John McPhee mentioned a hole "visible from the moon" in his book *Encounters with the Archdruid* (37–38), Brower had objected. The *Wall Street Journal* seized on McPhee's characterization that the claim exaggerated the facts. According to Brower's biographer, Tom Turner, the ad was not Brower's or the Sierra Club's but was prepared by Jerry Mander and run by the North Cascades Conservation Council, which seems to square with the archival traces. See Turner, *David Brower*, 158–59, 274n15.
49. See Box 209 and 210, HMJP, for correspondence related to North Cascades National Park, including copies of these coupons.
50. M. Brock Evans, to Dick Miller, Pittsburgh, October 10, 1967, Box 22, Folder 40, BEP.
51. Brock Evans, Memo to the North Cascades 1968 File, January 11, 1973, Box 10, folder: North Cascades 1972, SCNO.
52. Schulte, *Wayne Aspinall*, 227–34 (broad context), 232 ("molasses-moving"); Kaufman, *Henry M. Jackson*, 167. I benefited from conversations and correspondence in the spring 2017 with Steve Schulte about Aspinall's legislative style.
53. "Aspinall's Snit," *Seattle Post-Intelligencer*, April 23, 1968; Walt Woodward, "Many Barred from North Cascades Hearing," *Seattle Times*, April 19, 1968.
54. *Hearings before the Subcommittee on National Parks and Recreation*, 450–51; Sommarstrom, "Wild Land Preservation Crisis,"136 ("do his darndest").
55. Manning and North Cascades Conservation Council, *Wilderness Alps*, 194–95.
56. "Aspinall's Snit."
57. Quoted in Manning and North Cascades Conservation Council, *Wilderness Alps*, 195.
58. Manning and North Cascades Conservation Council, *Wilderness Alps*, 196.
59. Johnson, "Remarks upon Signing Four Bills Relating to Conservation."
60. Marsh, *Drawing Lines*, 58–60.
61. Louter, *Windshield Wilderness*, 130–33.
62. Wyss, *Man Who Built the Sierra Club*, 256.
63. Harvey Manning, Issaquah, Washington, to Henry M. Jackson, Washington, D.C., October 18, 1968, Box 209, Folder 27, HMJP. Emphasis in original.

CHAPTER 9. THE STUDENT

1. Rome, *Genius of Earth Day*, 37.
2. Benjamin Shaine, Oberlin, Ohio, to Brock Evans, Seattle, May 15, 1968, Box 22, Folder 36, BEP.
3. Ben[jamin Shaine], Aspen, Colorado, to Brock and Rachel [Evans], January 20, 1969, Box 22, Folder 36, BEP.
4. Shaine to Evans, May 15, 1968.
5. The sixties and its political movements are explored in many books and essays. A good general overview is Lytle, *America's Uncivil Wars*. For the student movement, see Breines, "Of This Generation." For all things sixties, see Farber and Bailey, *Columbia Guide to America in the 1960s*.
6. Benjamin A. Shaine, "The Proposed Open-Pit Copper Mine in the Glacier Peak Wilderness of Washington State: An Analysis," Oberlin College, May 1968, Box 22, Folder 46, BEP.
7. Gottlieb, *Forcing the Spring*, 132–34; Marcuse, *One-Dimensional Man*.
8. Commoner, *Closing Circle*, 33–46 (four laws of ecology),140–77 (technology). Egan, *Barry Commoner and the Science of Survival*, is the best and most sophisticated book on Commoner<b. Personal communication with Egan helped me shape this analysis.
9. Ehrlich, *Population Bomb*. Sabin, *Bet*, offers an incisive analysis of Ehrlich's contributions to the environmental movement, especially his role in polarizing environmental politics.
10. For analysis of the dispute, see Egan, *Barry Commoner*, 116–31.
11. Shaine, "Proposed Open-Pit Copper Mine," 5.
12. In "Fight for Wilderness Preservation in the Pacific Northwest," Brock Evans discusses how early activists (like him) did not have the biological language yet so relied on the language of wilderness and beauty. Evans, "Fight for Wilderness."
13. Shaine to Evans, May 15, 1968.
14. To reconstruct the taping of Douglas, see Benjamin A. Shaine, "Kennecott on Miners Ridge: A Description and Analysis of the Controversy over the Proposal of the Kennecott Copper Corporation to Construct an Open-Pit Copper Mine in the Glacier Peak Wilderness Area of Washington," honors thesis, Oberlin College, 1969; Harvey Manning, Issaquah, Washington, to Dave Brower, San Francisco, May 18, 1968; remarks of Justice William O. Douglas regarding Kennecott Copper Corporation's proposed open-pit copper mine on Miners Ridge, July 21, 1968; Points to Consider When Taping Remarks W. O. D., July 21, 1968, Goose Prairie, Washington: all in Box 8, Folder: Kennecott (Mine) Film 1968, N3CR.
15. A copy of the thesis draft is housed in the N3C's records. It runs about one hundred pages and contains notes for more. Shaine, "Kennecott on Miners Ridge."
16. Ben [Shaine], Oberlin, Ohio, to Brock Evans, Seattle, November 18, 1968, Box 22, Folder 36, BEP.
17. Shaine, "Kennecott on Miners Ridge."
18. Early on, Shaine inquired of Evans the legal strategy to be employed, willing for the details to be "off the record." Shaine to Evans, November 18, 1968.

19. Shaine, "Kennecott on Miners Ridge." Shaine also briefly investigated legislative solutions, such as reforming the Wilderness Act. He observed that the politics remained quite similar to 1964, so reducing the mining industry's power legislatively seemed unlikely.
20. Brooks, *Before Earth Day*, 149; Brosnan, "Law and the Environment," 536; Fox, *American Conservation Movement*, 304.
21. Ben [Shaine], Oberlin, Ohio, to Brock [Evans], May 16, 1969; Box 22, Folder 36, BEP; Shaine, "Kennecott on Miners Ridge" ("aggressive"); Rome, *Genius of Earth Day*, 191 ("bold"), 194 ("brash"); Gottlieb, *Forcing the Spring*, 189 ("combative").
22. Fox, *American Conservation Movement*, 304–5; Gottlieb, *Forcing the Spring*, 188–93; Rome, *Genius of Earth Day*, 190–97.
23. Victor John Yannacone Jr., Patchogue, N.Y., to Ben Shaine, Oberlin, Ohio, May 13, 1969; Vic Y., handwritten note presumably to Brock Evans, June 10, 1969, Box 22, Folder 39, BEP.
24. For Evans's interest in this strategy, see M. Brock Evans to Ben Shaine, Oberlin, Ohio, April 30, 1969, Box 22, Folder 40, BEP.
25. Yannocone to Shaine, May 13, 1969.
26. Shaine, "Kennecott on Miners Ridge."
27. Sam Roberts, "David M. Gates Dies at 94; Sounded Early Alarm on Environmental Perils," *New York Times*, March 11, 2016.
28. Sabin, *Bet*.
29. Egan explores this well in *Barry Commoner and the Science of Survival*, esp. 47–78.
30. Shaine to Evans, May 15, 1968.
31. Ben [Shaine] to Brock and Rachel Evans, January 20, 1969.
32. Ben [Shaine], Oberlin, Ohio, to Brock Evans, Seattle, March 4, 1969, Box 22, Folder 36, BEP.
33. Shaine, "Continuing Threat to Miners Ridge," 2–6 ("plush executive suite" on p. 2); Shaine, "Kennecott on Miners Ridge" ("polite, reasoned, self-assured").
34. Benjamin Shaine, Oberlin, to Frank R. Milliken, New York, April 11, 1969, Box 8, Folder: Kennecott Mine General Correspondence 1968–73, N3CR.
35. Ben [Shaine], New York, to Brock Evans, Seattle, April 11, 1969, Box 22, Folder 36, BEP ("top brass").
36. Nader's role in sparking public interest research groups around the country is mentioned in several books but still needs serious scholarly investigation. See Fox, *American Conservation Movement*, 305–6; Gottlieb, *Forcing the Spring*, 176; Hays, *Beauty, Health, and Permanence*, 460–61.
37. Shaine to Milliken, April 11, 1969.
38. Notes on April 11, 1969, meeting with Kennecott on Miners Ridge operation, Box 22, Folder 36, BEP. Shaine published a cleaned-up version of these notes in Shaine, "Continuing Threat to Miners Ridge," 2–6.
39. Notes on meeting, although most quotations are reproduced in both Shaine, "Kennecott on Miners Ridge," and Shaine, "Continuing Threat to Miners Ridge."
40. Shaine to Evans, April 11, 1969.

41. Notes on meeting, emphasis in original; Shaine to Evans, April 11, 1969.
42. Ben [Shaine], Oberlin, Ohio, to the Goldsworthys, April 10, 1969; Conservation Press Release, April 11, 1969: both in Box 8, Folder: Kennecott Mine General Correspondence 1968–73, N3CR.
43. Shaine to Evans, November 18, 1968; Ben [Shaine], Aspen, Colorado, to Brock and Rachel [Evans], January 20, 1969, Box 22, Folder 36, BEP.
44. "Imperiled Wilderness," *Washington Post*, May 15, 1969.
45. Benjamin Shaine, Oberlin, Ohio, to Lloyd Meeds, Washington, D.C., April 28, 1969; Lloyd Meeds, Washington, D.C., to Edward Weinberg, Washington, D.C., May 16, 1969; and David [illegible], Washington, D.C., to Lloyd Meeds, Washington, D.C., June 19, 1969: all in Box 8, Folder: Kennecott Mine General Correspondence 1968–73, N3CR.
46. Notes on meeting, emphasis in original.

CHAPTER 10. THE TROUBLE

1. Hyde, *Copper for America*, 200.
2. Shaine, "Continuing Threat to Miners Ridge," 2–6, quotation from 3.
3. Edelstein, "Copper," 46.
4. Charles B. Camp, "Down with Copper: Red-Metal Users Turn to Substitutes Even as Prices Plummet," *Wall Street Journal*, August 24, 1966.
5. Edelstein, "Copper," 45–46, quotation from 45.
6. Hyde, *Copper for America,* 190–94; Edelstein, "Copper," 45.
7. Clark H. Jones, San Bernardino, California, to Frank R. Milliken, New York, May 22, 1967, Box 8, Folder: Kennecott Mine General Correspondence 1967, N3CR.
8. Frank R. Milliken, New York, to Clark H. Jones, San Bernardino, California, June 12, 1967, Box 8, Folder: Kennecott Mine General Correspondence 1967, N3CR.
9. Milliken to Jones, June 12, 1967.
10. Milliken to Jones, June 12, 1967; Jake Booher, "Nuclear Shots Seen as Industry Tool," *Farmington (NM) Daily Times*, September 20, 1967 ("solution would be flushed"); "Atomic Explosion Experiment Asked as Method of Mining Copper Deposits," *Daily Journal of Commerce*, October 30, 1967; "Kennecott Proposes Nuclear Blast to Test Use as Aid in Mining," *Wall Street Journal*, October 13, 1967; "Nuclear Blast to Be Tried in Mining," unsourced newspaper clippingin Box 8, Kennecott Mine—Clippings 1966–1970, n.d., N3CR. See also Joralemon, *Copper*, 348–49.
11. Carr Childers, *Size of the Risk*, 141–45. I thank Leisl Carr Childers for this reference and forcing me to think through this point.
12. Milliken to Jones, June 12, 1967.
13. Jones to Milliken, May 22, 1967.
14. Milliken to Jones, June 12, 1967.
15. "3-Month Copper Strike Ended by Chilean Miners," *New York Times*, April 1, 1966.
16. Hyde, *Copper for America*, 202.

17. United Steelworkers of America, "Brief Union Summary."
18. "Copper Workers Strike Kennecott," *New York Times*, July 15, 1967; Robert Walker, "Copper Strike Is Taken Calmly," *New York Times*, August 13, 1967 ("stockpile heavily"); Robert Walker, "Storm of Apathy Greets Copper Rise," *New York Times*, February 18, 1968 (a chart shows world prices dropping just as the strike started, followed by a rise); "Break Reported in Copper Strike; 2 Companies Near Agreements," *New York Times*, March 11, 1968.
19. United Steelworkers of America, "Brief Union Summary."
20. "Figures in Strike against U.S. Copper Producers," *New York Times*, August 13, 1967.
21. "Copper Strike Improvisation," *New York Times*, January 26, 1968.
22. Edelstein, "Copper."
23. "Break Reported in Copper Strike"; Hyde, *Copper in America*, 202.
24. Camp, "Down with Copper"; Ray Vicker, "Four Major Copper-Producing Countries Meet Today on Price Levels, Cut in Output," *Wall Street Journal*, June 1, 1967. For "received in February 1967" and notations on clippings, see article copies in Box 8, Kennecott Mine—Clippings 1966–1970, n.d., N3CR. The copper industry predictably fought back, with an ad purchased in the *Wall Street Journal* stating, "A substitute for copper is exactly that. A substitute." *Wall Street Journal*, January 31, 1967.
25. "Rise in Copper Output to Offset Higher Costs Urged by Anaconda Co.," *Wall Street Journal*, September 7, 1967. This article also suggested that copper ought to be priced in a way to "discourage substitution of new metals or materials and to promote new uses of the red metal."
26. Richard L. Miller, Pittsburgh, to Brock Evans, Seattle, December 21, 1967, Box 22, Folder 39, BEP; Brock Evans, to Dick Miller, Pittsburgh, October 10, 1967, Box 22, Folder 40, BEP.
27. "Kennecott vs. the North Cascades," *Steel Labor*, January 1968, 7. In his study of labor in the Northwest timber industry, Erik Loomis noted several ways unions promoted healthier environments. See Loomis, *Empire of Timber*.
28. Draft article included with Miller to Evans, December 21, 1967.
29. M. Brock Evans, to Richard Miller, Pittsburgh, January 26, 1968, Box 22, Folder 40, BEP.
30. See letters from the Amalgamated Transit Union (registering their "strong opposition") and the Window Cleaners Union ("stating strongly our disapproval") as well as Jackson's reply to the Culinary Workers and Bartenders Union; August Antonino, Seattle, to Henry Jackson, Washington, D.C., April 19, 1968 (Transit Union); Hugh Sugiura, Seattle, to Henry M. Jackson, Washington, D.C., April 15, 1968 (Window Cleaners): all in Box 210, Folder 1, HMJP; Henry M. Jackson to Hazel M. Leirdahl and Harold A. Dodson, Renton, Washington, May 16, 1968, Box 210, Folder 5, HMJP.
31. Hyde, *Copper for America*, 197; "Kennecott Copper Corp. and Peabody Coal Co.," *New York Times*, July 8, 1966; "Kennecott Copper and Peabody Coal," *New York Times*, January 27, 1968; Eileen Shanahan, "Antitrust Violation Charged by F.T.C. against Kennecott," *New York Times*, August 13, 1968; "Kennecott

Request on Divesting Denied," *New York Times*, October 15, 1975; Herbert Koshetz, "Kennecott Sets Terms for Sale of Peabody Unit," *New York Times*, December 1, 1976.

32. A strike at a Kennecott subsidiary in 1966 shaped this yearly comparison, because it limited production. "Kennecott Profit Showed Big Gain in First Quarter," *Wall Street Journal*, May 3, 1967.
33. Robert Walker, "Kennecott Lists a Profit Decline," *New York Times*, February 6, 1968.
34. "Kennecott Lists a Profit Decline," *New York Times*, February 6, 1968; "Reported in Copper Strike."
35. "Kennecott Copper Profits Rise," *New York Times*, October 24, 1968.
36. Camp, "Down with Copper"; Vicker, "Four Major Copper-Producing Countries Meet Today"; "Rise in Copper Output to Offset Higher Costs Urged by Anaconda Co.," *Wall Street Journal*, September 7, 1967; "Extension of Remarks," *Congressional Record*, June 4, 1969, 4598–99; Box 22, Folder 46, BEP.

CHAPTER 11. THE STORIES

1. Brower worked on behalf of the North Cascades going back to the 1950s, but the Sierra Club's Northwest representatives, especially Brock Evans, did most of the work in the region for the club, always cooperating and coordinating with activists and organizations whose primary loyalty was regional.
2. "Our Wilderness Alps." This wasn't the first time *Sunset* featured the region. An article in 1958, "Skyscraper Country," guided readers to the "fabulous" North Cascades as the wilderness campaign warmed up.
3. "Our Wilderness Alps."
4. Brooks, *House of Life*.
5. Brooks, *Pursuit of Wilderness*, xi–xiii.
6. Brooks, "Fight for America's Alps," 87–99, quotations from 87, 89. This article reached an even wider audience when *Reader's Digest* reprinted a version of it three months later.
7. Brooks, *Pursuit of Wilderness*, 33.
8. Runte, *National Parks*, 156–57; Sellars, *Preserving Nature*, 183–87, 205–7.
9. Brooks, *Pursuit of Wilderness*, 42, 43.
10. Brooks, *Pursuit of Wilderness*, 3–4.
11. Paul Brooks, Boston, to Henry M. Jackson, Washington, D.C., March 29, 1967; Henry M. Jackson, to Paul Brooks, Boston, April 24, 1967; Henry M. Jackson, Washington, D.C., to Paul Brooks, Boston, May 11, 1967; Paul Brooks, Boston, MA, to Henry M. Jackson, Washington, D.C., June 21, 1967; Paul Brooks, Boston, to Henry M. Jackson, Washington, D.C., June 30, 1967: all in Box 209, Folder 27, HMJP. Brooks also checked with Brock Evans to ensure accuracy. See Evans's response, M. Brock Evans, to Paul Brooks, Boston, June 30, 1967, Box 22, Folder 40, BEP.
12. Brooks, *Pursuit of Wilderness*, 48, 49. The Library of America gathered some of Galbraith's most influential writings of this era in Galbraith, *Affluent Society and Other Writings*, quotation from 617.

13. Brooks, *Pursuit of Wilderness*, 55. Keith Makoto Woodhouse explores the rise and criticism of environmentalism through economics in *Ecocentrists*, chap. 4.
14. Richardson, *Bomb in Every Issue*, 1 (quotation), 57, 164.
15. Marine, *America the Raped.*
16. Richardson, *Bomb in Every Issue.*
17. Marine, *America the Raped*, 62, 61.
18. Marine, *America the Raped*, 72.
19. A useful way to see how science, conservation biology in particular, played a role in wilderness management discussions is found in Turner, "Conservation Science and Forest Service Policy."
20. Marine, *America the Raped*, 15.
21. An excellent overview of opposition to the "land skinners" is found in Miller, "Sylvan Prospect."
22. Marine, *America the Raped*, 16, 18.
23. Marine, *America the Raped*, 246; M. Brock Evans, to Gene Marine, San Francisco, October 10, 1967, Box 22, Folder 40, BEP.
24. Marine, *America the Raped*, 66 ("sneaky"), 65 ("ugly scar").
25. Marine, *America the Raped*, 65–66.
26. McPhee, *Encounters with the Archdruid.*
27. McPhee, "Profiles."
28. Park was a last-minute replacement for Henry Burgess of Kennecott, who was ordered by the company president to back out. Turner, *David Brower*, 156.
29. McPhee, *Encounters with the Archdruid*, 3–5, 9, quotations from 5, 9.
30. Richard White points to loggers making a moral claim for their daily connection to nature in "Are You an Environmentalist?"
31. McPhee, *Encounters with the Archdruid*, 14.
32. Park, *Affluence in Jeopardy*, v (quotation), 335–36.
33. McPhee, *Encounters with the Archdruid*, 17.
34. McPhee, *Encounters with the Archdruid*, 6 (intruders), 8 (thimbleberries), 9 (Douglas fir).
35. McPhee, *Encounters with the Archdruid*, 10, 19–20.
36. McPhee, *Encounters with the Archdruid*, 16–17, 21, 51, 41, 22, 23.
37. McPhee, *Encounters with the Archdruid*, 25.
38. McPhee, *Encounters with the Archdruid*, 24.
39. McPhee, *Encounters with the Archdruid*, 25.
40. USDA, Forest Service, Okanogan-Wenatchee National Forest, "Holden Mine Site Cleanup."
41. McPhee, *Encounters with the Archdruid*, 26.
42. Spring and Manning, *100 Hikes in Washington's North Cascades*, 51.
43. McPhee, *Encounters with the Archdruid*, 58, 59.
44. McPhee, *Encounters with the Archdruid*, 60, 42 (Brower and Park agreeing on population issues. See Sabin, *Bet*, for an insightful analysis of how population interacted with other issues of the day to animate debate and help polarize environmental issues. See also Woodhouse, *Ecocentrists*, for how population questions manifested and divided in radical environmentalist circles.

45. McPhee, *Encounters with the Archdruid*, 60, 63.
46. McPhee, *Encounters with the Archdruid*, 61 (emphasis in original), 74.
47. McPhee, *Encounters with the Archdruid*, 71, 72–73.
48. Turner, *David Brower*, 131–52 (quotation from resignation speech from 149 and "outgrown" from 151. See also Wyss, *Man Who Built the Sierra Club*, 242–72; McPhee, *Encounters with the Archdruid*, 208–20.
49. McPhee, *Encounters with the Archdruid*, 216.
50. Turner, *David Brower*, chaps. 11–12.

CONCLUSION

1. "Wilderness Mines: Cause for Concern," *Wild Cascades*, Spring 1977, quotations from 12.
2. Danner, *Crown Jewel Wilderness*, 209; Joel Connelly, "The Open Pit Is Finally Put Away," *Seattle Post-Intelligencer*, May 5, 2010, http://blog.seattlepi.com/seattlepolitics/2010/05/05/the-open-pit-is-finally-put-away (quotation); Bill Sheets, "Swap Adds Miner's Ridge to Glacier Peak," May 7, 2010, http://www.heraldnet.com/article/20100507/NEWS01/705079879; Consolidated Natural Resources Act of 2008, Public Law 110-229, *US Statutes at Large* 122 (2008), 758–59. Gary Paull helped me understand the conservation easement through personal correspondence in June 2017.
3. Manning and North Cascades Conservation Council, *Wilderness Alps*, 189, shares Meeds's perspective. Others also wondered about this at various points in the campaign.
4. Leshy, *Mining Law*, 235.
5. Journalist Joel Connelly is the only one I know to have advanced this exact argument. As an astute and long-standing observer of Northwest politics, Connelly is credible, and Jackson had the power to have followed through on such a threat. Yet such a claim is not possible to verify now. See Connelly, "Open Pit Is Finally Put Away." In her study of North Cascades National Park, Lauren Danner reports a similar claim about Jackson, although one less specific about using the 1872 Mining Law. Danner, *Crown Jewel Wilderness*, 209.
6. See, for instance, North Cascades Conservation Council, "Mining Issues," the N3C's own list of achievements.
7. Manning and North Cascades Conservation Council, *Wilderness Alps*, 189.
8. This scene is reconstructed from Brock Evans, "The Mining of Natural Beauty," *Cooperator: The Voice of the Puget Sound Cooperative League*, August 1966; Brock Evans, "Memo to Kennecott File," January 4, 1973, Box 22, Folder 44, BEP; and M. Brock Evans, Seattle, to Frank Milliken, New York, March 3, 1967, Box 22, Folder 40, BEP.
9. Evans, "Fight for Wilderness Preservation," 83, emphasis in original.
10. Debates over wilderness within environmental history and related fields have proliferated. An early volume that collected a first round of positions is found in Callicott and Nelson, eds., *Great New Wilderness Debate*. A recent collection that explores preservation in the context of the Anthropocene context is Minteer and

Pyne, *After Preservation*. A rousing multidisciplinary defense of wilderness in the context of these criticisms is Wuerthner, Crist, and Butler, *Keeping the Wild*.

11. Evans, "Fight for Wilderness Preservation," 84–86, quotation from 86.
12. Turner, *Promise of Wilderness*, 406.
13. Leshy, *Mining Law*, 235. The Cabinet Mountains in Montana have faced periodic mining company interest, however. See the excellent and detailed legal analysis of Loop, "Claiming the Cabinets."
14. Danner, *Crown Jewel Wilderness*, 190–92, 206–9; Marsh, *Drawing Lines*, 163n50.
15. M. Brock Evans, to Frank R. Millik[e]n, New York, July 18, 1967, Box 22, Folder 40, BEP. Emphasis in original.

BIBLIOGRAPHY

ARCHIVAL COLLECTIONS *(BY ABBREVIATION USED IN THE NOTES)*

BEP	Brock Evans Papers (accession #1776-006). Special Collections. University of Washington, Seattle.
GMP	George Marshall Papers. Bancroft Library, University of California, Berkeley.
HMJP	Henry M. Jackson Papers (accession #3560-04). Special Collections. University of Washington, Seattle.
N3CR	North Cascades Conservation Council Records (accession #1732-001). Special Collections. University of Washington, Seattle.
RUSFS	Records of the U.S. Forest Service. Record Group 95. National Archives, Seattle.
SCMP-EWP	Edgar Wayburn Papers. Sierra Club Members Papers. Bancroft Library, University of California, Berkeley.
SCNOR	Sierra Club, Northwest Office Records (accession #2678-001). Special Collections. University of Washington, Seattle.
WODP	William O. Douglas Papers. Library of Congress, Washington, D.C.
WSR	Wilderness Society Records. Conservation Collection. Denver Public Library.

GOVERNMENT DOCUMENTS

Hearing before the Committee on Agriculture and Forestry. U.S. Senate, 87th Cong., 1st sess., January 13, 1961. Washington, D.C.: Government Printing Office, 1961.

Hearings before the Committee on Interior and Insular Affairs, United States Senate, Study Team Report of the Recreational Opportunities in the State of Washington. 89th Cong., 2nd sess., February 11–12, 1966.

Hearings before the Subcommittee on National Parks and Recreation of the Committee on Interior and Insular Affairs, House of Representatives. 90th Cong., 2nd sess., on H.R. 8970 and related bills, April 19 and 20, 1968.

Hearings before the Subcommittee on Parks and Recreation of the United States Senate. 90th Cong., 1st sess., on S. 1321, April 24–25, May 25, 27, 29, 1967.
A National Plan for American Forestry. 73rd Cong., 1st sess., S. Res. 175. Washington, D.C.: Government Printing Office, 1933.
North Cascades Study Team. *The North Cascades Study Report: A Report to the Secretary of the Interior and the Secretary of Agriculture.* Washington, D.C.: U.S. Department of the Interior and U.S. Department of Agriculture, 1965.

PERIODICALS CITED

Bellingham Herald
Bremerton (WA) Sun
Columbian (Vancouver, WA)
Daily Journal of Commerce
Everett Herald
Farmington (NM) Daily Times
Los Angeles Times
Mountaineer (Seattle)
Newsweek
Oregonian (Portland)
New York Times
San Francisco Chronicle
Seattle Post-Intelligencer
Seattle Times
Skagit Valley Herald
Steel Labor
Wall Street Journal
Washington Post
Wenatchee (WA) Daily World
Wild Cascades

BOOKS, CHAPTERS, ARTICLES, AND WEBSITES

"About Kennecott: How Its Hirelings Spoke against the Glacier Peak Wilderness." *Wild Cascades*, December 1966–January 1967, 16.
Adams, Nigel B. *The Holden Mine: Discovery to Production, 1896–1938.* Wenatchee, Wash.: World Publishing Company for Washington State Historical Society, 1981.
Alinsky, Saul D. *Reveille for Radicals.* Chicago: University of Chicago Press, 1946. First published 1945.
———. *Rules for Radicals: A Practical Primer for Realistic Radicals.* New York: Random House, 1971.
Andrews, Thomas G. "'Made by Toile'? Tourism, Labor, and the Construction of the Colorado Landscape, 1858–1917." *Journal of American History* 92, no. 3 (December 2005): 837–63.
Arntz, Dee. *Extraordinary Women Conservationists of Washington.* Charleston, S.C.: History Press 2015.

Arrington, Leonard J., and Gary B. Hansen. *"The Richest Hole on Earth": A History of the Bingham Copper Mine.* Logan: Utah State University Press, 1963.

Basso, Keith H. *Wisdom Sits in Places: Landscape and Language among the Western Apache.* Albuquerque: University of New Mexico Press, 1996.

Beckey, Fred. *Range of Glaciers: The Exploration and Survey of the Northern Cascade Range.* Portland: Oregon Historical Society Press, 2003.

Bradley, Harold, with photographs by Ansel Adams. "The Northern Cascades: A Masterpiece to Preserve." *Sierra Club Bulletin* 43, no. 9 (November 1958): 28.

Braman, Jim. "Braman, James d'Orma (Dorm), (1901–1980)." HistoryLink.org, September 10, 2002, http://www.historylink.org/File/3919.

Breines, Wini. "'Of This Generation': The New Left and the Student Movement." In *Long Time Gone: Sixties America Then and Now,* edited by Alexander Bloom, 23–45. New York: Oxford University Press, 2001.

Brinkley, Douglas. *The Wilderness Warrior: Theodore Roosevelt and the Crusade for America.* New York: HarperCollins, 2009.

Brock, Emily K. *Money Trees: The Douglas Fir and American Forestry, 1900–1944.* Corvallis: Oregon State University Press, 2015.

Brooks, Karl Boyd. *Before Earth Day: The Origins of American Environmental Law, 1945–1970.* Lawrence: University Press of Kansas, 2009.

Brooks, Paul. "The Fight for America's Alps." *Atlantic Monthly*, February 1967, 87–99.

———. "The Fight for America's Alps." *Reader's Digest,* May 1967, 2–6.

———. *The House of Life: Rachel Carson at Work.* Boston: Houghton Mifflin, 1972.

———. *The Pursuit of Wilderness.* Boston: Houghton Mifflin, 1971.

Brosnan, Kathleen A. "Law and the Environment." In *The Oxford Handbook of Environmental History,* edited by Andrew C. Isenberg, 513–52. New York: Oxford University Press, 2014.

Brower, David. *For Earth's Sake: The Life and Times of David Brower.* Salt Lake City: Gibbs Smith, 1990.

Caldbick, John. "Frederick William Cleator (1883–1957), Forest Ranger." HistoryLink .org, January 30, 2012. http://www.historylink.org/File/10014.

Callicott, J. Baird, and Michael P. Nelson, eds. *The Great New Wilderness Debate.* Athens: University of Georgia Press, 1998.

Carr Childers, Leisl. *The Size of the Risk: Histories of Multiple Use in the Great Basin.* Norman: University of Oklahoma Press, 2015.

Chamberlin, Silas. *On the Trail: A History of American Hiking.* New Haven: Yale University Press, 2016.

Chriswell, Harold C. *Memoirs*. Bellingham, Wash.: H. Chriswell, 1989.

Cleator, Fred. "Report on Glacier Peak Wilderness Area, Washington State." Records of the U.S. Forest Service. Record Group 95, National Archives, Seattle.

Cliff, Edward P. "Mining and the National Forests." *Mining Congress Journal,* November 1969, 37–42.

———. "The Wilderness Act and the National Forests." In *Wilderness and the Quality of Life,* edited by Maxine E. McCloskey and James P. Gilligan, 6–12. San Francisco: Sierra Club, 1969.

Cohen, Lizabeth. *A Consumers' Republic: The Politics of Mass Consumption in Postwar America.* New York: Vintage, 2003.

Cohen, Michael P. *The History of the Sierra Club, 1892–1970.* San Francisco: Sierra Club Books, 1988.

Commoner, Barry. *The Closing Circle: Nature, Man, and Technology.* New York: Alfred A. Knopf, 1971.

Connelly, Joel. "North Cascades Conservation Council." HistoryLink.org, February 18, 2001, http://www.historylink.org/File/9714.

Cronon, William. *Nature's Metropolis: Chicago and the Great West.* New York: Norton, 1991.

———. "The Trouble with Wilderness, or, Getting Back to the Wrong Nature." In *Uncommon Ground: Rethinking the Human Place in Nature,* 69–90. New York: W. W. Norton, 1995.

———. "The Uses of Environmental History." *Environmental History Review* 17, no. 3 (Fall 1993): 1–22.

Danner, Lauren. *Crown Jewel Wilderness: Creating North Cascades National Park.* Pullman: Washington State University Press, 2017.

Darvill, Fred T., Jr. *Hiking the North Cascades.* Mechanicsburg, Penn.: Stackpole Books, 1998.

Dayton, Stan. "Behind the By-Lines." *Engineering and Mining Journal,* May 1967.

Dietrich, William. *The North Cascades: Finding Beauty and Renewal in the Wild Nearby.* Seattle: Braided River, 2014.

Dilsaver, Lary M., ed., *America's National Park System: The Critical Documents.* Lanham, Md.: Rowman & Littlefield, 1994. Digitized at National Park Service, https://www.nps.gov/parkhistory/online_books/anps/index.htm.

Douglas, William O. "America's Vanishing Wilderness." *Ladies' Home Journal,* July 1964, 37–41, 77.

———. *Farewell to Texas: A Vanishing Wilderness.* New York: McGraw-Hill, 1967.

———. *My Wilderness: East to Katahdin.* Garden City, NY: Doubleday, 1961.

———. *My Wilderness: The Pacific West.* Garden City, NY: Doubleday, 1960.

———. *Of Men and Mountains.* New York: Harper and Brothers, 1950.

Edelstein, Daniel. "Copper." In U.S. Geological Survey National Minerals Information Center, *Metal Prices in the United States through 2010.* https://minerals.usgs.gov/sir/2012/5188/sir2012-5188.pdf.

Edmund, Rudolph W. *Highlights of the Holden Copper Mine and its Geological Setting.* N.p: n.p., [1967].

Egan, Michael. *Barry Commoner and the Science of Survival: The Remaking of American Environmentalism.* Cambridge, Mass.: MIT Press, 2007.

Ehrlich, Paul. *The Population Bomb.* New York: Ballantine, 1968.

Evans, Brock. "The Fight for Wilderness Preservation in the Pacific Northwest." In *Protecting the Wild: Parks and Wilderness, the Foundation for Conservation,* edited by George Wuerthner, Eileen Crist, and Tom Butler, 83–92. Washington, D.C.: Island Press, 2015.

———. "The Mining of Natural Beauty." *Cooperator: The Voice of the Puget Sound Cooperative League,* August 1966, p. C.

Farber, David, and Beth Bailey, eds. *The Columbia Guide to America in the 1960s.* New York: Columbia University Press, 2001.

Farmer, Jared. *Glen Canyon Dammed: Inventing Lake Powell and the Canyon Country.* Tucson: University of Arizona Press, 1999.

Ficken, Robert E. *Rufus Woods, the Columbia River, and the Building of Modern Washington.* Pullman: Washington State University Press, 1995.

Fiege, Mark. *The Republic of Nature: An Environmental History of the United States.* Seattle: University of Washington Press, 2012.

Finks, P. David. *The Radical Vision of Saul Alinsky.* New York: Paulist Press, 1984.

Finney, Carolyn. *Black Faces, White Spaces: Reimagining the Relationship of African Americans to the Great Outdoors.* Chapel Hill: University of North Carolina Press, 2014.

Fox, Stephen. *The American Conservation Movement: John Muir and His Legacy.* Madison: University of Wisconsin Press, 1981.

Freeman, Orville L. "Address." In *Wilderness and the Quality of Life,* edited by Maxine E. McCloskey and James P. Gilligan, 107–15. San Francisco: Sierra Club, 1969.

Galbraith, James K., ed. *The Affluent Society and Other Writings, 1952–1967.* New York: Library of America, 2010.

Gates, Paul W. *History of Public Land Law Development.* Washington, D.C.: Government Printing Office, 1968.

General Services Administration. "A Brief History of GSA." http://www.gsa.gov/portal/content/103369.

Gilbert, Lewis D. *Dividends and Democracy.* Larchmont, N.Y.: American Research Council, 1956.

Glover, James M. *A Wilderness Original: The Life of Bob Marshall.* Seattle: Mountaineers, 1986.

Goldsworthy, Patrick D. "Kennecott Meets with Conservation Leaders." *Wild Cascades,* December 1966–January 1967, 4, 21.

Gottlieb, Robert. *Forcing the Spring: The Transformation of the American Environmental Movement.* Revised and updated ed. Washington, D.C.: Island Press, 2005.

Grauer, John Jack. "Mazamas." *Oregon Encyclopedia.* http://www.oregonencyclopedia.org/articles/mazamas/#.V40Ej7grI2w.

Gutierrez, David G. *Walls and Mirrors: Mexican Americans, Mexican Immigrants, and the Politics of Ethnicity.* Berkeley: University of California Press, 1995.

Hammond, John H., Jr. "The Wilderness Act and Mining: Some Proposals for Conservation." *Oregon Law Review* 47 (June 1968): 447–59.

Harvey, Mark W. T. *A Symbol of Wilderness: Echo Park and the American Conservation Movement.* Seattle: University of Washington Press, 2000. First published 1994.

———. *Wilderness Forever: Howard Zahniser and the Path to the Wilderness Act.* Seattle: University of Washington Press, 2005.

Hawley, Charles Caldwell. *A Kennecott Story: Three Mines, Four Men, and One Hundred Years, 1887–1997.* Salt Lake City: University of Utah Press, 2014.

Hays, Samuel P. *The American People and the National Forests: The First Century of the U.S. Forest Service.* Pittsburgh: University of Pittsburgh Press, 2009.

Hays, Samuel P., with Barbara D. Hays. *Beauty, Health, and Permanence: Environmental Politics in the United States, 1955–1985.* New York: Cambridge University Press, 1987.

Hirt, Paul W. *A Conspiracy of Optimism: Management of the National Forests since World War Two.* Lincoln: University of Nebraska Press, 1994.

Horwitt, Sanford D. *Let Them Call Me Rebel: Saul Alinsky—His Life and Legacy.* New York: Knopf, 1989.

Hurst, James Willard. *Law and the Condition of Freedom in the Nineteenth-Century United States.* Madison: University of Wisconsin Press, 1956.

Hyde, Charles K. *Copper for America: The United States Copper Industry from Colonial Times to the 1990s.* Tucson: University of Arizona Press, 1998.

Jagodinsky, Katrina. *Legal Codes and Talking Trees: Indigenous Women's Sovereignty in the Sonoran and Puget Sound Borderlands, 1854–1946.* New Haven: Yale University Press, 2016.

Johnson, Lyndon B. "Remarks upon Signing Four Bills Relating to Conservation and Outdoor Recreation." October 2, 1968. American Presidency Project. http://www.presidency.ucsb.edu/ node/237295.

———. "Special Message to Congress: Protecting Our Natural Heritage." January 30, 1967. American Presidency Project. http://www.presidency.ucsb.edu/node/237471.

Joralemon, Ira B. *Copper: The Encompassing Story of Mankind's First Metal.* Berkeley: Howell-North Books, 1973.

Kaufman, Herbert. *The Forest Ranger: A Study in Administrative Behavior.* Special reprint ed. Washington, D.C.: Resources for the Future, 2006.

Kaufman, Robert G. *Henry M. Jackson: A Life in Politics.* Seattle: University of Washington Press, 2000.

Kjeldsen, Jim. *The Mountaineers: A History.* Seattle: Mountaineers, 1998.

Klingle, Matthew W. *Emerald City: An Environmental History.* New Haven: Yale University Press, 2007.

Klyza, Christopher McGrory. *Who Controls Public Lands? Mining, Forestry, and Grazing Policies, 1870–1990.* Chapel Hill: University of North Carolina Press, 1996.

LeCain, Timothy J. *Mass Destruction: The Men and Giant Mines that Wired America and Scarred the Planet.* New Brunswick, N.J.: Rutgers University Press, 2009.

———. *The Matter of History: How Things Create the Past.* New York: Cambridge University Press, 2017.

Leopold, Aldo. "The Wilderness and Its Place in Forest Recreational Policy." In *The River of the Mother of God and Other Essays,* edited by Susan L. Flader and J. Baird Callicott, 78–81. Madison: University of Wisconsin Press, 1991.

Leshy, John D. *The Mining Law: A Study in Perpetual Motion.* Washington, D.C.: Resources for the Future, 1987.

Lewis, James G. *The Forest Service and the Greatest Good: A Centennial History.* Durham, N.C.: Forest History Society, 2005.

Loomis, Erik. *Empire of Timber: Labor Unions and the Pacific Northwest Forests.* New York: Cambridge University Press, 2016.

Loop, Donna J. "Claiming the Cabinets: The Right to Mine in Wilderness Areas." *Public Land Law Review* 7 (1986): 45–78.

Louter, David. *Windshield Wilderness: Cars, Roads, and Nature in Washington's National Parks.* Seattle: University of Washington Press, 2006.

Lytle, Mark Hamilton. *America's Uncivil Wars: The Sixties Era from Elvis to the Fall of Richard Nixon.* New York: Oxford University Press, 2006.

Maher, Neil M. *Nature's New Deal: The Civilian Conservation Corps and the Roots of the American Environmental Movement.* New York: Oxford University Press, 2008.

Manning, Harvey. *The Wild Cascades: Forgotten Parkland.* San Francisco: Sierra Club, 1965.

Manning, Harvey, with the North Cascades Conservation Council. *Wilderness Alps: Conservation and Conflict in Washington's North Cascades.* Bellingham, Wash.: Northwest Wild Books, 2007.

Marcuse, Herbert. *One-Dimensional Man: Studies in the Ideology of Advanced Industrial Society.* Boston: Beacon Press, 1991. First published 1964.

Marine, Gene. *America the Raped: The Engineering Mentality and the Devastation of a Continent.* New York: Simon and Schuster, 1969.

Marsh, Kevin R. *Drawing Lines in the Forest: Creating Wilderness Areas in the Pacific Northwest.* Seattle: University of Washington Press, 2007.

———. "The Ups and Downs of Mountain Life: Historical Patterns of Adaptation in the Cascade Mountains." *Western Historical Quarterly* 35, no. 2 (Summer 2004): 193–213.

[Marshall, Robert]. "Three Great Western Wildernesses: What Must be Done to Save Them?" *Living Wilderness* 1 (September 1935): 9–11.

McCloskey, Maxine E., and James P. Gilligan, eds. *Wilderness and the Quality of Life.* San Francisco: Sierra Club, 1969.

McCloskey, Michael. "Can Recreational Conservationists Provide for a Mining Industry?" *Rocky Mountain Mineral Law Institute* 13 (1967): 65–85.

———. *In the Thick of It: My Life in the Sierra Club.* Washington, D.C.: Island Press, 2005.

———. "Wilderness Movement at the Crossroads, 1945–1970." *Pacific Historical Review* 41, no. 3 (August 1972): 346–61.

McConnell, Grant. "The Cascade Range." In *The Cascades: Mountains of the Pacific Northwest,* ed. by Roderick Peattie, 65–96. New York: Vanguard Press, 1949.

———. "The Cascades Wilderness." *Sierra Club Bulletin* 41, no. 10 (December 1956): 24–31.

———. "Conservation and Politics in the North Cascades." In *Sierra Club Nationwide,* vol. 1, 1983. Oral history conducted by Rod Holmgren. http://digitalassets.lib.berkeley.edu/roho/ucb/text/sc_nationwide1.pdf.

———. "The Conservation Movement—Past and Present." *Western Political Quarterly* 7, no. 3 (September 1954): 463–78.

———. *Private Power and American Democracy.* New York: Knopf, 1966.

———. *Stehekin: A Valley in Time.* Seattle: Mountaineers Books, 2014. First published 1988.

McPhee, John. *Encounters with the Archdruid.* New York: Noonday Press, 1971.

———. "Profiles: C. Park and D. Brower." *New Yorker*, March 20, 1971, 42.
Merrill, Karen R. *Public Lands and Political Meaning: Ranchers, the Government, and the Property between Them.* Berkeley: University of California Press, 2002.
Miller, Char. *Gifford Pinchot and the Making of Modern Environmentalism.* Washington, D.C.: Island Press, 2001.
———. *Public Lands, Public Debates: A Century of Controversy.* Corvallis: Oregon State University Press, 2012.
———. "A Sylvan Prospect: John Muir, Gifford Pinchot, and Early Twentieth-Century Conservationism." In *American Wilderness: A New History,* edited by Michael L. Lewis, 131–48. New York: Oxford University Press, 2007.
Minteer, Ben A., and Stephen J. Pyne, eds. *After Preservation: Saving American Nature in the Age of Humans.* Chicago: University of Chicago Press, 2015.
Muir, John. *The Yosemite.* New York: Century, 1912.
Murphy, Bruce Allen. *Wild Bill: The Legend and Life of William O. Douglas.* New York: Random House, 2003.
Nash, Roderick Frazier. *Wilderness and the American Mind.* Fifth ed. New Haven: Yale University Press, 2014.
Needham, Andrew. *Power Lines: Phoenix and the Making of the Modern Southwest.* Princeton: Princeton University Press, 2014.
Neil, J. M. *To the White Clouds: Idaho's Conservation Saga, 1900–1970.* Pullman: Washington State University Press, 2005.
Neuberger, Richard L. *They Never Go Back to Pocatello: The Selected Essays of Richard Neuberger.* Edited by Steve Neal. Portland: Oregon Historical Society Press, 1988.
Nicolson, Marjorie Hope. *Mountain Gloom and Mountain Glory: The Development of the Aesthetics of the Infinite.* Seattle: University of Washington Press, 1997. First published 1959.
North Cascades Conservation Council. "Mining Issues." http://www.northcascades.org/wordpress/mining.
Nye, David E. *America as Second Creation: Technology and Narratives of New Beginnings.* Cambridge, Mass.: MIT Press, 2003.
"Our Greatest Wilderness Park-Land." *Sierra Club Bulletin*, June 1957.
"Our Wilderness Alps." *Sunset*, June 1965.
Park, Charles F., Jr., with Margaret C. Freeman. *Affluence in Jeopardy: Minerals and the Political Economy.* San Francisco: Freeman, Cooper, 1968.
Pearson, Byron. *Still the Wild River Runs: Congress, the Sierra Club, and the Fight to Save Grand Canyon.* Tucson: University of Arizona Press, 2002.
Peyton, Jonathan. *Unbuilt Environments: Tracing Postwar Development in Northwest British Columbia.* Vancouver: University of British Columbia Press, 2017.
Pinchot, Gifford. *Breaking New Ground.* Commemorative edition. Washington, D.C.: Island Press, 1998.
Poehlman, Elizabeth S. *Darrington: Mining Town/Timber Town.* Shoreline, Wash.: Gold Hill Press, 1995. First published 1979.
Pyle, Robert Michael. "Reflections on 50 Years of Engagement with the Natural World: Interview with Robert Michael Pyle." *Terrain.org: A Journal of the Built*

+ *Natural Environments*, May 31, 2015. http://www.terrain.org/2015/interviews/robert-michael-pyle.

Reich, Charles A. *The Greening of America*. New York: Random House, 1970.

Richardson, Peter. *A Bomb in Every Issue: How the Short, Unruly Life of* Ramparts *Magazine Changed America*. New York: New Press, 2009.

Rio Tinto. http://www.holdenminecleanup.com/.

Rogers, Jedediah S. *Roads in the Wilderness: Conflict in Canyon Country*. Salt Lake City: University of Utah Press, 2013.

Rome, Adam. *The Genius of Earth Day: How a 1970 Teach-In Unexpectedly Made the First Green Generation*. New York: Hill and Wang, 2013.

Rothman, Hal. *America's National Monuments: The Politics of Preservation*. Lawrence: University Press of Kansas, 1989.

———. "'A Regular Ding-Dong Fight': Agency Culture and Evolution in the NPS-USFS Dispute, 1916–1937." *Western Historical Quarterly* 20, no. 2 (May 1989): 141–61.

Runte, Alfred. *National Parks: The American Experience*. 4th ed. Lanham, Md.: Taylor Trade, 2010.

Sabin, Paul. *The Bet: Paul Ehrlich, Julian Simon, and Our Gamble over the Earth's Future*. New Haven: Yale University Press, 2013.

Schrepfer, Susan R. *The Fight to Save the Redwoods: A History of Environmental Reform, 1917–1978*. Madison: University of Wisconsin Press, 1983.

———. *Nature's Altars: Mountains, Gender, and American Environmentalism*. Lawrence: University Press of Kansas, 2005.

Schulte, Steven C. *Wayne Aspinall and the Shaping of the American West*. Boulder: University Press of Colorado, 2002.

Scott, Doug. *The Enduring Wilderness: Protecting Our Natural Heritage through the Wilderness Act*. Golden, Colo.: Fulcrum, 2004.

Sellars, Richard West. *Preserving Nature in the National Parks: A History*. New Haven: Yale University Press, 1997.

Shaffer, Marguerite S. *See America First: Tourism and National Identity, 1880–1940*. Washington, D.C.: Smithsonian Books, 2001.

Shaine, Ben. "The Continuing Threat to Miners Ridge." *Wild Cascades*, February–March 1970, 2–6.

Shorris, Earl. *Latinos: A Biography of the People*. New York: W. W. Norton, 1992.

Sierra Club. "Wilderness Alps of Stehekin." https://www.youtube.com/watch?v=Jc_EOcj6ZYw&.

Simon, James F. *Independent Journey: The Life of William O. Douglas*. New York: Harper & Row, 1980.

"Skyscraper Country." *Sunset*, August 1958, 43–48.

Sommarstrom, Allan Ralph. "Wild Land Preservation Crisis: The North Cascades Controversy." PhD diss., University of Washington, 1970.

Sowards, Adam M. "Administrative Trials, Environmental Consequences, and the Use of History in Arizona's Tonto National Forest, 1926–1996." *Western Historical Quarterly* 31, no. 2 (Summer 2000): 189–214.

———. *The Environmental Justice: William O. Douglas and American Conservation.* Corvallis: Oregon State University Press, 2009.

———. "Spiritual Egalitarianism: John Muir's Religious Environmentalism." In *John Muir in Historical Perspective,* edited by Sally M. Miller, 123–36. New York: Peter Lang, 1999.

Spring, Ira, and Harvey Manning. *100 Hikes in Washington's North Cascades: Glacier Peak Region.* Seattle: Mountaineers, 1988.

Steen, Harold K. *The U.S. Forest Service: A History.* Centennial ed. Durham, N.C.: Forest History Society; and Seattle: University of Washington Press, 2004.

[Stone, J. Herbert.] "Glacier Peak Wilderness Proposal: Statement on the Proposed Glacier Park [*sic*] Wilderness Area." *Western Camping World,* April 1960, 22–25.

Sutter, Paul S. *Driven Wild: How the Fight against Automobiles Launched the Modern Wilderness Movement.* Seattle: University of Washington Press, 2002.

Tabor, Rowland, and Ralph Haugerud. *Geology of the North Cascades: A Mountain Mosaic.* Seattle: Mountaineers, 1999.

Turner, James Morton. "Conservation Science and Forest Service Policy for Roadless Areas." *Conservation Biology* 20, no. 3 (June 2006): 713–22.

———. "From Woodcraft to 'Leave No Trace': Wilderness, Consumerism, and Environmentalism in Twentieth-Century America." *Environmental History* 7, no. 3 (July 2002): 462–84.

———. *The Promise of Wilderness: American Environmental Politics since 1964.* Seattle: University of Washington Press, 2012.

Turner, Tom. *David Brower: The Making of the Environmental Movement.* Oakland: University of California Press, 2015.

Unger, Nancy C. *Beyond Nature's Housekeepers: American Women in Environmental History.* New York: Oxford University Press, 2012.

United Steelworkers of America. "Brief Union Summary of the 1967 Copper Strike." September 25, 1967. http://digitalassets.lib.berkeley.edu/irle/ucb/text/lb001083.pdf.

US Department of Agriculture, Forest Service, Pacific Northwest Region. *Glacier Peak Land Management Study.* Portland, Ore.: US Department of Agriculture, Forest Service, Pacific Northwest Region, 1957.

———. *Glacier Peak Wilderness Proposal.* Portland, Ore.: US Department of Agriculture, Forest Service, Pacific Northwest Region, 1959.

US Department of Agriculture, Forest Service, Okanogan-Wenatchee National Forest. "Holden Mine Site Cleanup." https://www.fs.usda.gov/detail/okawen/landmanagement/projects/?cid=fsbdev3_053632.

US Environmental Protection Agency. "Case Summary: EPA and Forest Service Issue UAO at the Holden Mine Site." https://www.epa.gov/enforcement/case-summary-epa-and-forest-service-issue-uao-holden-mine-site.

Walter R. Skinner's "Mining Year Book," 1966. London: Vintry House, 1967.

Walter R. Skinner's Mining Year Book, 1969. London: Vintry House, 1970.

Warren, Louis S. *The Hunter's Game: Poachers and Conservationists in Twentieth-Century America.* New Haven: Yale University Press, 1997.

Warth, John F. "The Glacier Peak Wilderness." *National Parks Magazine*, October–December 1956, 173–76, 193–94.
Watkins, T. H. *Righteous Pilgrim: The Life and Times of Harold L. Ickes, 1874–1952.* New York: Henry Holt, 1990.
White, Richard. "'Are You an Environmentalist, or Do You Work for a Living?': Work and Nature." *Uncommon Ground: Toward Reinventing Nature,* edited by William Cronon, 171–85. New York: W. W. Norton, 1995.
———. "Contested Terrain: The Business of Land in the American West." In *Land in the American West: Private Claims and the Common Good,* ed. by William G. Robbins and James C. Foster, 190–206. Seattle: University of Washington Press, 2000.
Wilkinson, Charles F. *Crossing the Next Meridian: Land, Water, and the Future of the West.* Washington, D.C.: Island Press, 1992.
Williams, Gerald W. *The U.S. Forest Service in the Pacific Northwest: A History.* Corvallis: Oregon State University Press, 2009.
Wolfe, Linnie Marsh, ed. *John of the Mountains: The Unpublished Journals of John Muir.* Madison: University of Wisconsin Press, 1938.
Wood, Linda Sargent. *A More Perfect Union: Holistic Worldviews and the Transformation of American Culture after World War II.* New York: Oxford University Press, 2012.
Woodhouse, Keith Makoto. *The Ecocentrists: A History of Radical Environmentalism.* New York: Columbia University Press, 2018.
Woodhouse, Philip R., with Robert L. Wood. *Monte Cristo.* Seattle: Mountaineers, 1996. First published 1979.
Wuerthner, George, Eileen Crist, and Tom Butler, eds. *Keeping the Wild: Against the Domestication of Earth.* Washington, D.C.: Island Press, 2014.
Wyckoff, William, and Lary M. Dilsaver. "Defining the Mountainous West." In *The Mountainous West: Explorations in Historical Geography,* edited by William Wyckoff and Lary M. Dilsaver, 1–59. Lincoln: University of Nebraska Press, 1995.
Wyss, Robert. *The Man Who Built the Sierra Club: A Life of David Brower.* New York: Columbia University Press, 2016.
Young, Terence. *Heading Out: A History of American Camping.* Ithaca: Cornell University Press, 2017.
Zalesky, Philip H. "Glacier Peak Area: Wilderness or Waste?" *Mountaineer* 48, no. 13 (December 1955): 37–40.

INDEX

References to illustrations appear in italic type.

CPSIA information can be obtained
at www.ICGtesting.com
Printed in the USA
LVHW021158190620
658367LV00011B/170